PEARL ANNIVERSARY EDITION

THE MONTAUK PROJECT: EXPERIMENTS IN TIME was originally released in 1992, causing an uproar and shocking the scientific, academic, and journalistic communities, all of whom were very slow to catch on to the secret world that lurks beyond the superficial veneer of American civilization.

A COLLOQUIAL NAME FOR SECRET EXPERIMENTS that took place at Montauk Point's Camp Hero, the Montauk Project represented the apex of extensive research carried on after World War II; and, in particular, as a result of the phenomena encountered during the Philadelphia Experiment of 1943 when the United States Navy attempted to achieve radar invisibility.

THE MONTAUK PROJECT ATTEMPTED TO STUDY why and how human beings, when exposed to high powered electromagnetic waves, suffered mental disorientation, physical dissolution or even death. A further ramification of this phenomena is that such electromagnetic waves rescrambled components of the material universe itself. According to reports, this research not only included successful attempts to manipulate matter and energy but also time itself.

IT HAS NOW BEEN A FULL THIRTY YEARS SINCE *The Montauk Project* originally appeared in print. In this Pearl Anniversary Edition, you will not only read the original text but also accompanying commentary by Peter Moon which includes details that could not be published at the original time of publication. Finally, you will receive news that the subsequent investigation of the Montauk Project culminated in the scientific proof of time travel capabilities. Turn the page and discover the most exalted capabilities of Man.

▲

OTHER TITLES FROM SKY BOOKS

by Preston Nichols and Peter Moon
The Montauk Project: Experiments in Time
Montauk Revisited: Adventures in Synchronicity
Pyramids of Montauk: Explorations in Consciousness
Encounter in the Pleiades: An Inside Look at UFOs
The Music of Time

by Peter Moon
The Black Sun: Montauk's Nazi-Tibetan Connection
Synchronicity and the Seventh Seal
The Montauk Book of the Dead
The Montauk Book of the Living
Spandau Mystery

by Radu Cinamar with Peter Moon
Transylvanian Sunrise
Transylvanian Moonrise
Mystery of Egypt — The First Tunnel
The Secret Parchment
Inside the Earth — The Second Tunnel
Forgotten Genesis
The Etheric Crystal

by Stewart Swerdlow
Montauk: The Alien Connection
The Healer's Handbook: A Journey Into Hyperspace

by Alexandra Bruce
The Philadelphia Experiment Murder:
Parallel Universes and the Physics of Insanity

by Wade Gordon
The Brookhaven Connection

EXPERIMENTS IN TIME

PEARL ANNIVERSARY EDITION

PRESTON B. NICHOLS
AND PETER MOON

SkyBooks

NEW YORK

The Montauk Project : Experiments in Time
PEARL ANNIVERSARY EDITION
Copyright © 1992 by Preston B. Nichols and Peter Moon
(Copyright renewal applied for)
Additional notations Copyright © 2022 by Peter Moon
First printing (original edition), June 1992
Fifteenth printing (original edition), October 2018
First printing of Pearl Anniversary Edition, August 2022

Cover art, Illustrations, typography and design:
Creative Circle Inc.
Editorial Consultant, Odette de La Tour
Published by: Sky Books
                Box 769
                Westbury, New York 11590
                www.skybooksusa.com
                www.timetraveleducationcenter.com
                email: skybooks@yahoo.com

DISCLAIMER The nature of this book necessitates clear statements of what is and what is not being purported. This story is based upon the memory, recollections and experiences of Preston Nichols. He has recounted these events to the best of his ability. It is up to the reader to evaluate their relative truth. The publisher does not assume responsibility for inaccuracies that may have resulted from induced trauma or misconceptions. Many names and locations have been withheld or changed to protect the privacy of those concerned. Lastly, nothing in this book should be interpreted to be an attack on the United States Government. The publisher and the authors believe and fully support the United States Government as set forth by the U.S. Constitution. The heinous activities described herein are considered to by perpetrated by individuals who were not acting within the legal bounds of the law.

Library of Congress Cataloging-in-Publication Data

Nichols, Preston B. / Moon, Peter
       The Montauk Project: Experiments in Time
       Pearl Anniversary Edition
by Preston B. Nichols and Peter Moon
       268 pages, illustrated
       ISBN 978-0-9631889-0-8 (13 digit)
1. Occult Science 2. Time travel 3. Anomalies
Library of Congress Catalog Card Number 2018963494

This book is dedicated to Dr. David Lewis Anderson, the founder of the Time Travel Research Center, the World Genesis Foundation, and the Anderson Institute. It is also dedicated to all the personnel of those organizations, including the citizens of Romania, who have supported his work as well as that of Peter Moon.

ACKNOWLEDGEMENTS

Keith Allen
Charlene Babb
Marion Berrian
Bob Beutlich
Al Bielek
Duncan Cameron
Jeff Cave
Odette de la Tour
George R. Dickson
John Ford &
Long Island UFO Network
Margaret Geiger
Dr. Fred Goldrich
Claude Hensley
Betty Hughes
Judith Pope Koteen
Howard Metz
John Odin
Dillon Ridguard
Clarence Robbins
Lorraine Saluzzi
Dr. Mel Sobol
Stewart Swerdlow
U.S. Psychotronics Association
And countless others who shall
remain nameless

TABLE OF CONTENTS

INTRODUCTION

It is now thirty years since the original version of *The Montauk Project* was released to the public in 1992. Many sequels and supplementary information have followed, including the revelation that time travel capability can be proven within the context of ordinary math and physics. (This is now easily explained at the Time Travel Education Center (see *www.timetraveleducationcenter.com*).

What follows is the original text with additional italicized annotations by myself, most of which were not considered to be appropriate for public consumption at the time of the original publication. These are no longer issues as the two major players who originally broke the Montauk Project story, Preston Nichols (1945-2018) and Duncan Cameron (1951-2019), have passed away.

Original Introduction

At the eastern most end of Long Island sits Montauk Point, known to most New Yorkers for its scenic beauty and landmark lighthouse. To the immediate west of the lighthouse, there is a mysterious and derelict Air Force base on the grounds of old Fort Hero. Although it was officially decommissioned and abandoned by the U.S. Air Force in 1969, it was subsequently reopened and continued to operate without the sanction of the U.S. Government.

The entire financing for the base is also a mystery. No funding can be traced to the military or government. Officials of the U.S. Government have probed for answers without success.

The secrecy of the operation has prompted legends to thrive across Long Island. However, it is unlikely that any of the local people of Montauk, or those who spread the tales, know the full story of what actually went on there.

A circle of insiders believe the Montauk Project was a development and culmination of the phenomena encountered aboard the *USS Eldridge* in 1943. Popularly known as the "Philadelphia Experiment", the ship actually disappeared while the Navy conducted radar invisibility experiments.

According to these accounts, over three decades of secret research and applied technology ensued. Experiments were conducted that included electronic mind surveillance and the control of distinct populations. The climax of this work was reached at Montauk Point in 1983. It was at that point that the Montauk Project effectively ripped open a hole in space-time to 1943.

Perhaps the person best qualified to tell the real story is Preston Nichols, an electrical engineer and inventor who has studied the Montauk Project for the better part of a decade. His interest in the project was spurred in part by unusual circumstances in his own life. He was also able to legally acquire much of the equipment that was used for the project. His continued investigation ultimately revealed his own role as the technical director of the project. Despite brainwashing and threats to silence him, he has survived and has decided it is in the best interest of all to tell his story.

GUIDE TO THE READER

Because the subject matter of this book is controversial, we would like to offer some guidelines.

This book is an exercise in consciousness. It is an invitation to view time in a new manner and expand your awareness of the universe. Time rules our fate and ushers in our death. Although we are regulated by its laws, there is much that we do not know about time and how it relates to our consciousness. Hopefully, at the very least, this information will broaden your horizons.

Some of the data you will read in this book can be considered as "soft facts". Soft facts are not untrue, they are just not backed up by irrefutable documentation. A "hard fact" would be documentation or hard physical evidence that could stand up to scrutiny.

By the nature of the subject matter and security considerations, hard facts about the Montauk Project have been very difficult to obtain. There is also an area between soft and hard which can be termed "gray facts". These would be very plausible but not as easily provable as a hard fact.

Any serious investigation will show that a Montauk Project did, in fact, exist. One can also find people who have been experimented on in some fashion or another.

This book is not an attempt to prove anything. The purpose is to get a story told that is of essential interest to scientific researchers, metaphysicians and citizens of the planet Earth. It is the story of one particular individual and his circle of contacts. It is hoped that more individuals will come out of the closet and that researchers will come forth with more investigations and documentation.

This work is being presented as non-fiction as it contains no falsehoods to the best knowledge of the authors. However, it can also be read as pure science fiction if that is more suitable to the reader.

A short glossary has been provided in the back to assist with ordinary electronic terms and those of a more esoteric nature. Scientists who read this book should understand that the definitions are designed to assist the general reader's understanding. They are not purported to be the latest technical jargon. Likewise, the general reader should understand that the diagrams in this book are included for the benefit of technical people. If one is interested, they can get a further understanding of those terms and symbols by studying the *Radio Amateur's Handbook* or a text of a similar nature.

NOTE: *The above guide is as applicable today as it was when it was written. I would add, however, that Preston Nichols has always tried to avoid hard facts as stated above, not because they do not exist but rather because, per his statement, he is afraid to prove the Montauk Project for fear of his life. While I believe his fears are passe due to the time that has passed since his employment in the defense industry, it does not behoove myself nor anyone else to overlook the trauma he experienced and how this might affect his decision making. As the technology of time travel has improved considerably (as you will later read) since the days of the Montauk Project, I would also consider this to be a mitigating factor with regard to disclosure of hard facts.*

1 THE PHILADELPHIA EXPERIMENT

The origin of the Montauk Project dates back to 1943 when radar invisibility was being researched aboard the *USS Eldridge*. As the *Eldridge* was stationed at the Philadelphia Navy Yard, the events concerning the ship have commonly been referred to as the "Philadelphia Experiment". Having been the subject of different books and a movie, only a quick synopsis will be given here.*

The Philadelphia Experiment was known as the Rainbow Project to those who manned and operated it. It was designed as a top secret project that would help end World War II. The forerunner of today's stealth technology, the Rainbow Project was experimenting with a technique to make a ship invisible to enemy radar. This was done by creating an "electromagnetic bottle" which actually diverted radar waves around the ship. An "electromagnetic bottle" changes the entire electromagnetic field of a specific area — in this case, the field encompassing the *USS Eldridge*.

While the objective was to simply make the ship undetectable by radar, it had a totally unexpected and drastic side effect. The ship became invisible to the naked eye and left the space-time continuum. Soon thereafter, the ship suddenly reappeared in Norfolk, Virginia, hundreds of miles away.

The project was a success from a material standpoint, but it was a drastic catastrophe to the people involved.

---

* Further information on the Philadelphia Experiment can be found in Appendix E.

While the *USS Eldridge* "moved" from the Philadelphia Naval Yard to Norfolk and back again, the crew found themselves in complete disorientation. They had left the physical universe and had no familiar surroundings to relate to. Upon their return to the Philadelphia Navy Yard, some were planted into the bulkheads of the ship itself. Those who survived were in a mental state of disorientation and absolute horror.

The crew were subsequently discharged as "mentally unfit" after having spent considerable time in rehabilitation. The status of "mentally unfit" made it very convenient for their stories to be discredited.

This put the Rainbow Project at a standstill.

Although a major breakthrough had occurred, there was no certainty that human beings could survive further experimentation. It was too risky. Dr. John von Neumann, who headed the project, was now summoned to work on the Manhattan Project. This concerned the making of the atom bomb which became the weapon of choice for ending World War II.

Although it is not well known, vast research that began with the Rainbow Project was resumed in the late 1940s. It continued on, culminating with a hole being ripped through space-time at Montauk in 1983. The goal of this book is to give you a general understanding of the research and events subsequent to the Philadelphia Experiment and up to 1983 at Montauk. I will begin by telling you how I, Preston Nichols, stumbled across it.

EDITOR'S NOTE (from Peter Moon):

Preston's duties in the defense industry required him to read a detailed classified report on the Philadelphia Experiment. More will be said on this later as it coincides within the context of this book.

Following the publication of the original book, I received a report from Al Bielek that Preston Nichols had once been personally given a movie reel of original footage of the actual Philadelphia Experiment but had returned it to the Government. Al was rather outraged over this issue. When I asked Preston about it, he acknowledged that this was true, but he also stated that he was obligated to return the footage because it was not his to possess and would also violate certain security agreements he had signed in accordance with his job duties.

Years later, I learned from an entirely separate source that Preston had read a detailed report on the Philadelphia Experiment when he was working for AIL, a major defense contractor on Long Island whose full name is Airborne Instruments Laboratory. The report was detailed and technical in nature, and it was done in the course of his job duties. When asked, Preston also acknowledged that this was true. He could not, however, reveal further information due to non-disclosure agreements he had signed.

A M P L I T R O N

A key component of the Montauk Project was the amplitron.
Essentially a high powered UHF amplifier, the amplitron served
as the final amplifier of the transmitter before a function was
radiated out the antenna. A large tube, it weighed
300 pounds and measured 35 inches in
its largest dimension.

2 MONTAUK DISCOVERED

In 1971, I began working for BJM*, a well known defense contractor on Long Island. Through the years, I got a degree in electrical engineering and became a specialist in electromagnetic phenomena. I was not then aware of the Philadelphia Experiment or its accompanying phenomena.

Although I was not extraordinarily interested in the paranormal at that time, I had obtained a grant to study mental telepathy and to determine whether or not it existed. I sought to disprove it, but I was surprised to find out that it did, in fact, exist.

I began my research and found out that telepathic communication operated on principles that are strikingly similar to that of radio waves. I had discovered a wave that could be termed a "telepathic wave". In some respects, it behaved like a radio wave. I set out to get the characteristics of this "telepathic wave". I studied their wave lengths and other pertinent facts. I determined that while a telepathic wave behaves like a radio wave, it isn't exactly a radio wave. Although it propagates in a similar fashion to that of electromagnetic waves and possesses like properties, not all of these fit into normal wave functions.

I found all of this very exciting. I had discovered a whole brand new electromagnetic function that was not in any of the text books I'd ever seen. I wanted to learn

---

* BJM is a fictitious name for the company I worked for. (*NOTE: The actual company, we can safely inform you twenty-five years later, is AIL or Airborne Instruments Laboratory. More will be said on this in further notations.*)

as much as I could and studied all the activities that might use this type of function. My interest into metaphysics had been launched.

I continued to research in my spare time and collaborated with different psychics to test and monitor their various responses. In 1974, I noticed a peculiar phenomena that was common to all of the psychics that I worked with. Every day, at the same hour, their minds would be jammed. They couldn't think effectively. Suspecting that the interference was caused by an electronic signal, I used my radio equipment and correlated what came on over the air waves at the times the psychics were non-functional. Whenever a 410-420 MHz (Megahertz) cycle appeared on the air, they were jammed. When the 410-420 MHz cycle was off, the psychics would open back up after about twenty minutes. It was obvious that this signal was greatly impeding the ability of the psychics.

I decided to trace this signal. Placing a modified TV antenna on the roof of my car, I grabbed a VHF receiver and set out looking for the source of it. I tracked it right to Montauk Point. It was coming directly from a red and white radar antenna on the Air Force base.

At first, I thought that this signal might have been generated accidentally. I checked around and found out that the base was still active. Unfortunately, security was tight and the guards wouldn't give any useful information. They said that the radar was for a project run by the FAA. I couldn't press the point beyond that. In fact, their statement didn't make a lot of sense. This was a World War II radar defense system known as "SAGE Radar". It was totally antiquated, and there is not any known reason why the FAA would need such a system. I didn't believe them but couldn't help being intrigued. Unfortunately, I had hit a dead end.

I continued my psychic research, but didn't get anywhere on the investigation of the Montauk antenna until 1984 when a friend of mine called. He told me the place was now abandoned, and that I should go out there and check it out. I did. It was indeed abandoned with debris strewn everywhere. I saw a fire extinguisher left amidst many scattered papers. The gate was opened as were the windows and doors of the buildings. This is not the way the military normally leaves a base.

I strolled around. The first thing that caught my eye was the high voltage equipment. I was very interested as it was a radio engineer's delight. I am a collector of ham gear and radio equipment, and I wanted to buy it. I figured it would be available cheap if I made the proper arrangements through the Surplus Disposal Agency in Michigan.

After examining all the equipment, I contacted the disposal agency and spoke to a friendly lady. I told her what I wanted, and she told me she would see what could be done. It appeared to be abandoned material and looked like a scrap contract. If this was so, I'd be able to take what I wanted. Unfortunately, I didn't hear from her so I called her back three weeks later. She informed me that there had not been any success with tracing the equipment. They couldn't find out who owned it. Neither the military nor the GSA (General Services Administration) claimed to know anything about it. Fortunately, the Surplus Disposal Agency said they would continue to track the matter further. After another week or two went by, I called her back. She said she'd turn me over to a John Smith (fictitious name), located at a military overseas terminal in Bayonne, New Jersey.

"Talk to him and he'll set something up," she said. "We like to keep our customers satisfied."

I met John Smith. He didn't want to discuss anything

on the phone. He said that no one officially admitted to owning that equipment. As far they were concerned, the equipment was abandoned and I could go in and take whatever I wanted. He gave me a piece of paper which appeared to be official and said to show it to anyone who might question my presence in the area. It was not an official document nor was it registered with anybody, but he assured me that it would keep the police off my back. He also referred me to the caretaker of the Montauk Air Force Base who would show me around.

3 A VISIT TO MONTAUK

I was out at the base within the week. There I met the caretaker, Mr. Anderson. He was very helpful. He told me to be careful and showed me where things were so that I wouldn't fall through the floor and that type of thing. He said I was welcome to take anything I could this trip, but if he ever saw me out there again, he'd have to kick me out. His job, after all, was to keep people off the base. He realized that the permission I had was semi-official at best. He was also kind enough to tell me that he went out for a drink every evening at 7 PM.

I had taken the trip to Montauk with a fellow named Brian. Brian was a psychic who had helped me with my research. As we foraged around the base, we went in two different directions. I went into a building and saw a man who appeared to be homeless. He told me that he had been living in the building ever since the base was abandoned. He also said that there had been a big experiment a year earlier and that everything had gone crazy. Apparently, he'd never gotten over it himself.

In fact, the man recognized me, but I had no idea who he was or what he was talking about. I did listen to his story. He said he had been a technician at the base and that he'd been AWOL. He had deserted the project just before the base had been abandoned. He spoke about a big beast appearing and frightening everyone away. He told me a lot about the technical details of the machinery and how things worked. He also said something that was very strange. He told me that he remembered me well. In fact, I had been his boss on the project. Of course, I thought it was pure nonsense.

I didn't know then that there was any truth to his story. This was just the beginning of my discovery that the Montauk Project was real.

I left the man and found Brian. He was complaining that things weren't right and that he was feeling some very funny vibrations. I decided to ask him for a psychic reading right there. His reading was strangely similar to what the homeless man had just told me. He spoke of irregular weather patterns, mind control and a vicious beast. He mentioned animals being affected, crashing through windows. Mind control was a main focus of Brian's reading.

The reading was interesting, but we were there to cart out the equipment. Much of it was heavy and we weren't allowed to bring a vehicle right onto the base. We had to back pack it. I was thus able to acquire much of the equipment left behind from the Montauk Project.

A few weeks later, I was surprised by a visitor who barged into my lab. He came straight to the lab which was in back of the house. He didn't ring the door bell or anything. He claimed to know me and said that I had been his boss. He went on to explain many of the technical details of the Montauk Project. His story corroborated what psychics and the homeless man had told me. I didn't recognize him but listened to all he had to say.

I was sure that something had gone on at the Montauk base, but I didn't know what. My personal involvement was evident, but I still didn't consider it very seriously.

I was, however, puzzled by different people recognizing me. I had to make it my business to investigate Montauk. So, I went out and camped on the beach for a week or so. I went to bars and asked the locals for stories about the base. I talked to people on the beach, on the street, wherever I could find them. I asked all about the strange activities that were purported to have occurred.

Six different people said that it had snowed in the middle of August. There were listings of hurricane force winds that came out of nowhere. Thunder storms, lightning and hail were also reported under unusual circumstances. They would appear when previously there had been no meteorological evidence to expect such.

There were other unusual stories besides the weather. These included stories of animals coming into the town en masse and sometimes crashing through the windows. By this time, I had taken different psychics out to the base. The stories confirmed what psychics had been able to determine through their own sensitivity.

I finally got the idea to speak to the Chief of Police who also informed me of strange happenings. For example, crimes would be committed in a two hour period. Then, all of a sudden, nothing. Keep in mind that Montauk is a very small town. After the quiet, another two hour period of crimes would occur. Teens were also reported to suddenly group en masse for two hours, then mysteriously separate and go their own ways. The Chief couldn't account for it, but his statements lined up perfectly with what the psychics had indicated about mind control experiments. I had collected some really bizarre information, but I didn't have many answers. I was, however becoming very suspicious. I had often travelled to ham-fests (where ham radio equipment is bought and sold), and there more people would recognize me. I had no idea who they were, but I would talk to them and ask them about Montauk. As I did, more information came, but everything was still a big puzzle.

MONTAUK AIR FORCE BASE
An overview, looking north. The computer control center is to the right. Just behind that is an office building. The round building to the left is a radar building that was also used for storage.

EDITOR'S NOTE (from Peter Moon):

Like many who have listened to Preston Nichol's stories, I have always maintained a healthy skepticism as to whether or not they are completely factual. As Preston offered virtually no details about the character named Brian, I wondered whether or not he really existed. To my surprise, within a year of the publication of the book, I met a man at a book signing who actually knew Brian. He said, however, that Brian had moved to Michigan and wanted nothing to do with this subject.

While it is indeed healthy to remain skeptical, it is equally healthy to keep an open mind.

RADAR REFLECTOR

Above is the huge radar reflector that sits atop the transmitter
building at the Montauk Air Force Base. Nearly as long as a football
field, it was used in the early experiments to beam
mood control functions.

4 DUNCAN ARRIVES

In November of '84, another man appeared at my lab door. His name was Duncan Cameron. He had a piece of audio equipment, and he wanted to know if I could help him with it. He quickly became absorbed in the group of psychics I had working with me at the time. This endeavor was a continuation of my original line of research. Duncan showed a keen aptitude for such work and was extremely enthusiastic. I thought he was too good to be true and became suspicious of him. My assistant, Brian, felt the same. He didn't like Duncan's sudden involvement and decided to go his own way.

At one point, I surprised Duncan by telling him that I would be taking him some place to see if he recognized it. I drove him to the Montauk Air Force Base. He not only recognized it, he told me what the purpose was for each of the various buildings. He knew exactly where the bulletin board in the mess hall was and many other such minute details. Obviously, he had been there before. He knew the place like the back of his hand. He provided new information about the nature of the base and what his own function had been. Duncan's input dovetailed very nicely with the previous data I had collected.

When Duncan entered the transmitter building, he suddenly went into a trance and began spitting out information. This was curious, but I had to shake him repeatedly to break him out of it. When I brought him back to the lab, I applied techniques that I'd learned to help Duncan unblock his memories. Layers of programming were now coming out of Duncan. A lot of information concerned the Montauk Project.

Many different things were revealed until finally, a shocking program came straight to the awareness of Duncan's conscious mind. He blurted out that he had been programmed to come to my place, befriend me and then kill me and blow up the entire lab. All my work would be totally destroyed. Duncan appeared to be more outraged at all this than I was. He swore that he would no longer help those who had programmed him, and he has worked with me ever since.

Further work with Duncan revealed even more bizarre information. He had been involved in the Philadelphia Experiment! He said that he and his brother Edward had served aboard the *Eldridge* as members of the crew.*

A lot of things surfaced as a result of my work with Duncan. I started to remember things about the Montauk Project and was now certain I'd been involved. I just didn't know how or why. The puzzle was slowly clearing up. I found Duncan to be an extremely operational psychic and through him I was able to confirm new information.

---

*An account of Duncan's role in the Philadelphia Experiment is in the book *The Philadelphia Experiment & Other UFO Conspiracies* by Brad Steiger with Al Bielek and Sherry Hanson Steiger.

EDITOR'S NOTE (from Peter Moon):

Duncan was indeed a very real character. He was lecturing with Preston the very night I met him. Duncan also appeared at conferences from time to time. A somewhat public character, he was also reclusive.

5 A CONSPIRACY REVEALED

I visited Montauk many more times, often with different people who had been involved. A small group of us began to realize that we had stumbled across one of the highest security projects the country had ever known. We figured that we had better do something fast with this new found knowledge. If we didn't, we might end up dead.

As a group, we decided action had to be taken. We weren't sure exactly what to do, so we sat around and discussed it. What was the best thing to do? Publish it? Immediately? We talked about it extensively. In July of 1986, we decided that I should go to the United States Psychotronics Association (USPA) in Chicago and talk about it. I did, and it created an uproar. Word got around fast to those who didn't want the Montauk story to be revealed. Suddenly, here I was, giving an unannounced lecture. The information got out to hundreds of people, and it helped our safety considerably. We couldn't be swept under the rug without creating a public furor. To this day, I still appreciate the open forum and free speech that the USPA provided me.

Now, we decided to feed the information to the Federal Government. One of my associates knew the nephew of a senior senator from the Southwest. The nephew, who we will call Lenny, worked for the Senator. We gave the information to Lenny who passed it to his uncle. This information included pictures of the orders given to the different military personnel which we had found strewn about the base.

The Senator did a personal investigation and verified that military technicians had in fact been assigned to the base. The Senator also discovered that the base was decommissioned, derelict and mothballed since 1969. Having served his country as an Air Force general, he was particularly interested to know why Air Force personnel were working on a derelict base. And, where did the money come from to open up the base and run it?

After they did their own investigation and saw the pictures and documents we supplied them, there was no question that the base had been active. They verified that Fort Hero (which is the name of the original World War I base that surrounds the entire area of the Air Force base) and Montauk were indeed derelict and were simply listed as property held by the General Services Administration since 1970.

The Senator got very involved and travelled to Long Island to find out what he could about the Montauk Air Force Base. He was not greeted with enthusiastic cooperation despite having very impressive personal credentials. People reported seeing him looking through the fences and trying to find out what was going on. He visited me and told me to keep quiet about it as speaking out any further could jeopardize his investigation. That is why I have kept this story quiet until now.

When the Senator completed his investigation, he couldn't find any trace of government funding, no appropriations, no oversight committees and no payments. He eventually retired due to advancing age, but I have since been informed by Lenny that he sees no problem with my story being published. He also said that the Senator is still in the picture and that the investigation had been reopened.

BULLETIN

MONTAUK AFS, NY

FIRE
FIRE LOCATION INFORMATION
MEDICS (Duty hours)
ALTERNATIVES CRISIS LINE
Father Traynor
NUMBER: 3

Dial 6
242
283-4440
668-2200

CRIME STOP
SECURITY POLICE (Duty Hours)
MEDICS (Non-duty hours)
EMERGENCY REPORTING LINE
Reverend Delamain
AUXILIARY CHAPLAINS

200
213
274
668-2700
475-7406

O F F I C I A L

DATE: 22 January 1981

1. PRAYER MEETING: The prayer meeting for Hostages release which had been planned by Father Traynor will not be conducted. However, I feel sure that each member of Montauk Air Force Station and each member of the local community shares a common feeling of thankfulness over the conclusion of this issue. [CC/211]

2. CHECK CASHING FOR TDY PERSONNEL: If you are here TDY and unable to cash a check due to closure of BX and other station facilities, please contact the First Sergeant. [CC/211]

3. BARBER SERVICE: 26 January 1981 will be the last day the barber will visit the site. [CCF/294]

4. ATTENTION ALL PERSONNEL: Due to lack of participation, the commissary run to Mitchell Field on 24 Jan 81 has been cancelled. [LGT/262]

5. ATTENTION ALL PERSONNEL: Anyone who purchased items through the unit fund auction are advised that the items will be available for pick up on Friday, 23 Jan 81, between 0900 and 1100. Items must be paid for at the time of pick up. This will be the last chance for pick up of these items. [LGT/262]

6. SURF & TURF: Will be served on 23 January 1981 at the dining hall. This will be the last time that Surf & Turf will be served at the dining hall, so come to the dining hall on the 23rd of January (Friday) and enjoy it. [SVF/249]

7. BASE CLOSURE: The Riverhead radar site will become operational this morning at 0900. Montauk AFS operations section will have full operational responsibility until tomorrow afternoon. At that time it is planned for Montauk operations to go into standby status. Standby will be for the operations section and fire alerting procedures will continue normal mission support. Normal base operator and fire alerting procedures will continue. If no major problems are encountered in the operations at Riverhead, this "operations standby" status will continue until 1 Feb, when Montauk AFS will officially close for operational purposes. [CC/211]

OFFICIAL

LEON F. PEARSALL, MSgt, USAF
Chief of Administration

MILES O. MARTIN, Major, USAF
Commander

AIR FORCE ORDERS

The orders above and on the following pages were found strewn about the base during an authorized visit. They establish that the base was indeed active and included military personnel. Originally, we blanked out names on some of the documents to protect the privacy of the particular individuals concerned.

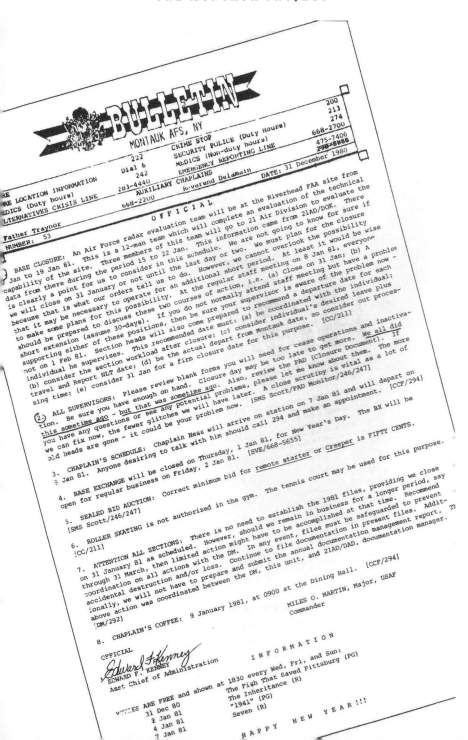

BULLETIN

MONTAUK AFS, NY

		200
		213
		274
CRIME STOP	222	
SECURITY POLICE (Duty Hours)	Dial 6	668-2700
MEDICS (Non-duty hours)	242	475-7406
EMERGENCY REPORTING LINE		298-5655
AUXILIARY CHAPLAINS	283-4440	
Reverend Delamain		DATE: 31 December 1980

RE LOCATION INFORMATION
RE (Duty hours)
EDICS (Duty hours)
LTERNATIVES CRISIS LINE   668-2200

Father Traynor
NUMBER: 53

O F F I C I A L

1. BASE CLOSURE: An Air Force radar evaluation team will be at the Riverhead FAA site from
Jan to 19 Jan 81. This is a 12-man team which will complete an evaluation of the technical
capability of the site. Three members of this team will go to 21 Air Division to evaluate the
data from there during the period 15 to 22 Jan. This information came from 21AD/DOK. There
is clearly a point for us to consider in this schedule. We are not going to know for sure if
we will close on 31 January or not until the last day or two. We must plan for the closure
because that is what our orders tell us to do. However, we cannot overlook the possibility
that it may be necessary to operate for an additional short period. At least it would be wise
to make some plans for this possibility. At the regular staff meeting on 8 Jan 81, everyone
should be prepared to discuss these two courses of action, i.e. (a) Close on 31 Jan; (b) A
short extension (assume 30-days). If you do not normally attend staff meeting but have a problem
supporting either of these positions, then be sure your supervisor is aware of the problem now -
not on 1 Feb 81. Section heads will also come prepared to recommend a departure date for each
individual he supervises. This recommended date must: (a) be coordinated with the individual;
(b) consider the section workload after closure; (c) consider individual's desired leave plus
travel and Report NLT date; (d) be the actual depart from Montauk date, so consider out proces-
sing time; (e) consider 31 Jan for a firm closure date for this purpose. [CC/211]

2. ALL SUPERVISORS: Please review blank forms you will need for cease operations and inactiva-
tion. Be sure you have enough on hand. Closure day may be too late to get more. We all did
this sometime ago - but that was sometime ago. Also, review the PAD (Closure Document). If
you have any questions or see any potential problems, please let me know about them. The more
we can fix now, the fewer glitches we will have later. A close scrutiny is vital as a lot of
old heads are gone - it could be your problem now. [SMS Scott/PAD Monitor/246/247]

3. CHAPLAIN'S SCHEDULE: Chaplain Hess will arrive on station on 7 Jan 81 and will depart on
9 Jan 81. Anyone desiring to talk with him should call 294 and make an appointment. [CCF/294]

4. BASE EXCHANGE will be closed on Thursday, 1 Jan 81, for New Year's Day. The BX will be
open for regular business on Friday, 2 Jan 81. [SVE/668-5655]

5. SEALED BID AUCTION: Correct minimum bid for remote starter or Creeper is FIFTY CENTS.
[SMS Scott/246/247]

6. ROLLER SKATING is not authorized in the gym. The tennis court may be used for this purpose.
[CC/211]

7. ATTENTION ALL SECTIONS: There is no need to establish the 1981 files, providing we close
on 31 January 81 as scheduled. However, should we remain in business for a longer period, say
through 31 March, then limited action might have to be accomplished at that time. Recommend
coordination on all actions with the DM. In any event, files must be safeguarded to prevent
accidental destruction and/or loss. Continue to file documentation in present files. Additionally, we will not have to prepare and submit the annual documentation management report. Th
above action was coordinated between the DM, this unit, and 21AD/DAD, documentation manager.
[DM/292]

8. CHAPLAIN'S COFFEE: 9 January 1981, at 0900 at the Dining Hall. [CCF/294]

MILES O. MARTIN, Major, USAF
Commander

OFFICIAL

Edward F. Kenney
EDWARD F. KENNEY
Asst Chief of Administration

I N F O R M A T I O N

MOVIES ARE FREE and shown at 1830 every Wed, Fri, and Sun:
31 Dec 80   The Fish That Saved Pittsburg (PG)
2 Jan 81   The Inheritance (R)
4 Jan 81   "1941" (PG)
7 Jan 81   Seven (R)

H A P P Y   N E W   Y E A R !!!

REQUEST AND AUTHORIZATION FOR CHANGE OF ADMINISTRATIVE ORDERS
(If more space is required, use reverse, identifying items by number)

TO: DAAO

FROM: DPMUO

TELEPHONE: 3217

THE FOLLOWING ORDER IS ☒ AMENDED AS SHOWN IN ITEM 5 ☐ RESCINDED ☐ REVOKED

1. IDENTIFICATION OF ORDER BEING CHANGED *(Issued by this Headquarters unless otherwise stated in item 6.)*

BASIC ORDER

A. PARA: —
B. ORDER *(Type and Number)*: SO AA-837
C. DATE: 20 MAY 80
D. TED: SEP 80

2. PREVIOUSLY AMENDED BY

A. PARA: —
B. ORDER *(Type and Number)*: SO AA-1896
C. DATE: 10 NOV 8

3. RELATING TO (TDY, PCS, SEPARATION, etc.)
PCS TO 966 AWACTS (TAC) TINKER AFB, OK 73145/CAFSC: 30554/RNLTD: 31 MAR 81/AAN:
0 900TN0177

☒ PCS WITH PCA (EDCSA)
☐ PCS WITHOUT PCA

4. IDENTIFICATION OF INDIVIDUALS TO WHOM CHANGE ACTION PERTAINS

A. GRADE: SGT
B. LAST NAME, FIRST, MIDDLE INITIAL: RONNIE A.
GONE
C. AFSN AND SSAN OR CIVILIAN POSITION TITLE: 280-64-4572
D. ORGANIZATION: 773 RADS (TAC)

5. AMENDMENT *(Identify item in order being amended)*

A. ITEM	AS READS	IS AMENDED TO READ
1	SRA	SGT
13	RONNIE A. 280-64-4572 / DET OL AA20 ADS (SAGE) (TAC)	966 AWACTS (TAC) RONNIE A. 280-64-4572
12	OCEANA/SOUCEK FLD VA 23460	TINKER AFB, OK 73145
37	1	9
	1 CSG/DPMUM LANGLEY AFB, VA 23665	2854 ABG/DPMUM TINKER AFB, OK 73145

B. ITEM: 5
IS AMENDED TO *(Include) (Delete)*
INCLUDE: MBR IS SCHEDULED TO ATTEND THE FOLLOWING CRS TDY ENROUTE PCS. CRS S-V86-A WATER SURVIVAL TRAINING COURSE AT HOMESTEAD AFB, FL, CLASS 81043, START IO MAR - GRAD 13 MAR 81. MBR WILL BE ATTACHED TO THE 3613 (SEE REVERSE)

6. REMARKS
5713500 321 5863.0* S503725
CIC:
5713500 321 5868.ON S503725
TAC:
TDY EXPENSE CHARGEABLE TO: 5713400 301 04A328 04A329 A8 S525002

7. DATE: 20 JAN 81

8. ORDERS ISSUING/APPROVING OFFICIAL *(Typed name, grade, and title)*
JERRY E. , TSGT, USAF
NCOIC, OUTBOUND ASSIGNMENTS

9. SIGNATURE

10. DESIGNATION AND LOCATION OF HEADQUARTERS
DEPARTMENT OF THE AIR FORCE
HQ 438TH MILITARY AIRLIFT WING (MAC)
MCGUIRE AIR FORCE BASE, NEW JERSEY 0864

11. ORDER *(Type and Number)*: SO AA-108

12. DATE: 20 Jan 81

13. TDN: FOR THE COMMANDER

14. DISTRIBUTION: "A"

15. SIGNATURE ELEMENT OF ORDERS AUTHENTICATING OFFICIAL
HEADQUARTERS 438TH MILITARY AIRLIFT WING OFFICIAL
WILLIAM , Captain, USAF
Chief, Central Base Administration

AF FORM JUL 70 973 PREVIOUS EDITION WILL BE USED

☆ U.S. GPO. 1975-0-567-954/62

REQUEST AND AUTHORIZATION FOR PERMANENT CHANGE OF STATION - MILITARY
(THIS FORM IS AFFECTED BY THE PRIVACY ACT OF 1974 - USE BLANKET PAS - AF FORM 11)

	2 PDAFSC/CAFSC	3 AIRMAN PAY GRADE	4 OVER 2 YEARS SERVICE (Yr mth)
	30554	E-3	

The following individual will proceed on permanent change of station
1 GRADE, LAST NAME, FIRST, MIDDLE INITIAL, SSAN

RONNIE A. 280-64-4572

5 TOY   AIC

GONE

6 REPORT TO COMDR, NEW ASSIGNMENT NLT   30 SEP 80

8 TED   SEPT 80   ☒PCS WITH PCA   ☐PCS WITHOUT PCA   1 DAYS

12 TPC WITH   TRAVEL TIME PERMITTED

11 LEAVE ADDRESS K134 HOCHWALT   DAYTON OHIO 45408

14 UNIT, MAJOR COMMAND, AND ADDRESS OF UNIT FROM WHICH RELIEVED
773 RADAR SQ (TAC)
MONTAUK AFS NY 11954

7 PDD   NA

10 ODALVF   YES

9 SECURITY CLEARANCE   SEC/ENTAC/JUL 77

15 UNIT OF UNIT TO WHICH ASSIGNED

13 UNIT, MAJOR COMMAND, AND ADDRESS OF UNIT TO WHICH ASSIGNED
DET OL AA 20 AIR DEF SAGE SQ (TAC)
OCEANA/SOUCEK FLD VA 23460

16 TRAVEL OF DEPENDENT(S) IS AUTHORIZED   ☐ACCOMPANIED TOUR   ☐ALL OTHERS TOUR   ☐DEPENDENT(S) PROHIBITED WITHIN OVERSEA AREA
17. AUTHORITY FOR CONCURRENT TRAVEL

18 VOLUNTEER STATUS   ☐VOLUNTEER   ☐NON-VOLUNTEER
☐VOLUNTEER ☐NON-VOLUNTEER

20. EXCESS BAGGAGE AUTHORIZED   POUNDS   PIECES

21. DISLOCATION ALLOWANCE CATEGORY   NA

19. INDIVIDUAL ELECTED TO SERVE
☐CONCURRENT ☐TO A DESIGNATED LOCATION
19 DEPENDENT(S) (List names of dependent(s) and DOB of children)

23. HOUSING AVAILABILITY (Oversea Assignment Only) (Check applicable block)
A. ☐ WILL BE AVAILABLE WITHIN 20 WEEKS. TRANSPORTATION OF DEPENDENTS AND SHIPMENT OF HHG AUTHORIZED TO A DESIGNATED LOCATION BUT WILL EXHAUST FURTHER TRAVEL AND TRANSPORTATION ENTITLEMENTS UNTIL MEMBER RECEIVES NEW PCS ORDERS. DEPENDENTS ARE AUTHORIZED SHIPMENT OF UNACCOMPANIED BAGGAGE TO A DESIGNATED LOCATION AND SUBSEQUENT SHIPMENT TO THE MEMBER'S OVERSEA DUTY STATION.
B. ☐ WILL NOT BE AVAILABLE WITHIN 20 WEEKS. TRANSPORTATION OF DEPENDENTS, SHIPMENT OF HHG AND UNACCOMPANIED BAGGAGE AUTHORIZED TO DESIGNATED LOCATION AND SUBSEQUENTLY TO MEMBER'S OVERSEA DUTY STATION.

22. OVERSEA TRANSPORTATION DATA:
A. ☐ COMPLY WITH MTA (DD Form 1482)
B. ☐ MEMBER WILL COMPLY WITH REPORTING TIME AND FLIGHT RESERVATIONS IN THE MTA OR AS ARRANGED BY THE TMO PER AFM 75-8, ATCH 1, AND IS NOT AUTH TO DEPART THIS STATION BEFORE RECEIPT OF VALIDATED MTA OR GTR (AF 1169) FROM THE TMO.
C. ☐ TDY STATION WILL OBTAIN FLIGHT RESERVATIONS. MEMBER IS NOT AUTH TO DEPART TDY STATION BEFORE RECEIPT OF VALIDATED MTA OR GTR FROM THE TMO.
D. ☐ DEPENDENT(S) WILL COMPLY WITH REPORTING DATA AND FLIGHT RESERVATIONS IN THE MTA.

(*Insert M, D, H, I, L, T, or Y)

24. PCS EXPENSE CHARGEABLE TO: 5703500 320 5863.04 S503725

TAC: 5703500 320 5868.0N S503725

29. AUTHORITY AND PCS CODE   G
AFR 39-11 PCS CODE
AAN: 0900TN0177

CIC:

25 TDY EXPENSE CHARGEABLE TO   NONTEMPORARY STORAGE CHARGEABLE TO

Pursuant to AFR 30-15, you will report to the base housing referral office servicing your new duty station before entering into any rental, lease, or purchase agreement for off-base housing. If TDY en route is authorized, attach receipts showing cost of all lodgings used.   Early reporting is authorized.

27. REMARKS (Submit travel voucher within 5 workdays after completion of travel.
Items 1 & 2 on reverse apply.

30. SIGNATURE   [signature]

31. DATE   20 May 80

28. TYPED NAME, GRADE, AND TELEPHONE NO. OF CBPO OFFICIAL
J.W.   , CMSGT, USAF, 3217

32. SPECIAL ORDER NO.   AA-837

29. DATE   14 MAY 80

31. DESIGNATION AND LOCATION OF HEADQUARTERS
DEPARTMENT OF THE AIR FORCE
HQ 438 MILITARY AIRLIFT WING (MAC)
McGUIRE AFB NJ 08641

34. TDN

FOR THE COMMANDER

35. SIGNATURE ELEMENT OF ORDERS AUTHENTICATING OFFICER

HARRY W.   , Ca
Chief, Central Base Adm

[stamp: HEADQUARTERS OFFICIAL USAF 438 MILITARY AIRLIFT WING]

36. DISTRIBUTION   "A"

37. ADDRESS OF GAINING CBPO
1 CSG/DPMUM LANGLEY AFB VA 23665

AF FORM 899 JUN 76   PREVIOUS EDITION IS OBSOLETE.

Editor's Note: (from Peter Moon)

As mentioned on page 29, the first public mention of the Montauk Project was in Chicago at the U.S. Pyschotronics Association. The word "psychotronics" was coined at a 1972 conference in Prague and refers to the interface between electronics and the human mind, body and spirit. The coining of psychotronics is credited to Ingo Swann, sometimes heralded as the "Father of Remote Viewing" who rose to notoriety from experiments at the Stanford Research Institute where he was actually documented to have moved a deeply buried and otherwise inaccessible magnetometer with his psi powers. Although he tried to distance himself from his association with Scientology, he was a dedicated Scientologist during that period and was more than enthusiastic about the subject, acting as a full blown participant in L. Ron Hubbard's counseling techniques.

This USPA lecture was attended by Eugenia Macer-Storey, a writer from New York, who personally witnessed Preston Nichols having a loud shouting argument with at least two men outside the lecture hall. They were threatening Preston not to talk about the Montauk Project, and he shouted over them, saying that he would. Preston figured that if he gave the talk and disappeared or was killed subsequently, it would draw too much attention to the subject. Had he said nothing, the public ramifications of eliminating him would have been far less severe.

The senator mentioned in this chapter refers to Barry Goldwater. His nephew, Lenny, had heard Al Bielek speaking about Montauk on the radio and contacted him. This initial information was fed Senator Goldwater who began his own private investigation. Goldwater was a general in the U.S. Air Force who had a public interest

*in UFOs. He is famous for recounting his conversation
with General Curtis LeMay who, after being asked about
the subject of UFOs, vehemently told the Senator never
to bring that subject up again.*

*Before this book was written, Preston used to talk
about Goldwater's excursion to Montauk Point where
he was said to be looking through the fence at this un-
explainable military installation that was supposedly a
state park but was inaccessible to the public and had received funding from an unknown and mysterious source.*

*At the time the Montauk Project was published, Pres-
ton was adamant about Goldwater's name not being mentioned, but it was common knowledge around the Long
Island chapter of the U.S. Psychotronics Association.*

*Although the documents shown here appear to be
innocuous, they were a point of contention with a member of the New York State Park Police who accosted me
during my first visit to Camp Hero. He had been alerted
to "visitors on the base" and was very angry that myself
and another had penetrated the fenced area. We had
actually found a path where the gate was down. The po-
liceman was very concerned about whatever documents
I had found. After he told us to leave and departed, I
stuffed them in my pocket and returned to my car that
was parked about a mile or so away.*

6 "PROJECT MOONBEAM"

While the Senator was searching for paper trails that might reveal the secrets at Montauk, I knew that they would not solve my personal mysteries one bit. I had been recognized by people I didn't know, and it was obvious that I had severe memory blocks. What made things so hard to reconcile was that I had a full set of "normal" memories which told me where I had been.

My memory improved while working with Duncan, and I eventually realized that I must have been existing on two separate time tracks. As bizarre as it may sound, it was the only sensible explanation under the circumstances.

As my memory was still largely blocked, there were three avenues of approach to the problem. First, I could simply try to remember the other time track, through regression or hypnosis. This proved to be very difficult for me and was virtually of no use. Secondly, I could look for clues and hints (in our normal time track) that the other time track did, in fact, exist. Thirdly, I could try to find the answers through technology. This would include theories of how the other time track was created and how I ended up on it.

The third approach was easiest. I am told that many people might find this very confusing, but I was familiar with the theories of the Philadelphia Experiment and was not intimidated by physics or electromagnetism. I found it plausible. The second approach also proved extremely helpful, but clues were hard to come by.

It was now 1989. I started to roam around the plant at BJM, where I was still working. I would talk to different people and dredge up what information I could without trying to appear suspicious. I would also walk around and just sense my own personal gut reaction to the different places in the plant.

I became particularly irritated when I would come to a certain room. My innards would just churn. I sensed very strongly that there was something in that room that was disturbing me. I had to investigate it. I rang the doorbell and was told that I couldn't come in. It was a high security area. Reportedly, only ten people at the plant had the proper clearance to be in that room.

I found that virtually no one knew anything about it. Finally, I did find two people who'd been in there, but they said they couldn't tell me anything. One of them must have turned me in because the security personnel visited me shortly thereafter. It was time to lay low for a while.

About a year after my futile investigation, the room was totally cleared out. The doors were open and anyone could walk right in. It was obvious that there had been all sorts of equipment. Dirt markings revealed that four round things had stood on the floor. I presumed they were coil structures. It was clear that there had been a console. There was also a huge power line that still ran into the room. The entire place gave me the creeps, but I was driven to find out all I could.

I discovered an elevator in the back of the room. I got in and found only two buttons: Main Floor and Sub Floor. There was also a numbered key pad. I pushed the button for Sub Floor and tried to go down, but the elevator would only go so far. I heard a voice that told me to punch in the proper coded numbers on the key pad. I didn't have the code and a beeping noise went off for about thirty seconds. Security was alerted. I had hit another dead end.

I wasn't scoring any points with security, and it was time to lay low once again. I began to think of how I could show that something very unusual was going on.

I also recalled earlier strange experiences that had occurred while working at BJM. There was a period when, all of a sudden, a band-aid would appear on my hand. It hadn't been there fifteen minutes ago! I couldn't remember putting it on. This happened more than a few times.

One day, I had been sitting at my desk and my hand suddenly started to ache. The back of the hand was sore, and there was a band aid on it. I absolutely knew that I had not put that band aid on nor had I had it put on. I became very suspicious. I got up and went down to the nurse.

I said to her, "This may sound wacky, but was I in here for a band-aid?"

"No, you weren't in here," she told me.

I asked her where I'd gotten it and she said, "You must've gotten it from one of the first aid kits. Don't you remember?"

"I'm just trying to figure it out," I said and walked out.

I thought in my mind, "I'm not going to get a band-aid at BJM except from the company nurse." I wanted a record so I made a conviction that I would never use a first aid kit.

I eventually remembered the reason I had sustained so many injuries to my hands. In my alternate reality, I frequently had to move different equipment. I was just about the only one who could move it as most people would go wacky when they'd get near it. For some reason, it didn't seem to bother me. But it was heavy and hard to maneuver. With no one to assist me, bruised hands and band-aids became a regular occurrence.

I kept to my conviction not to use any band-aids from first aid kits. I continued to check with the nurse when they appeared, and the records indicated I'd never been to her. As this was an irregularity, she must have reported it to security. They visited me and said, "Why are you asking about band-aids, Mr. Nichols?" I knew better than to pursue that anymore.

Recalling these experiences with the band-aids helped spur my memory back to 1978. I remembered sitting at my work bench one day. All of a sudden, I smelled the scent of burning transformers. It was pungent, like the smell of burning tar. It came and disappeared very fast. This happened at 9:00 o'clock in the morning. The rest of the day continued as normal until 4:00 o'clock in the afternoon when the whole plant began to smell like putrid smoke from burning transformers.

I thought to myself, "That's the same smell I smelled at 9:00 o'clock this morning". But now it occurred to me that the event probably hadn't happened at the time I had thought. You can't burn up a transformer and have the smell disappear as fast as it had that morning.

Many more events of this nature had occurred. Each puzzle tended to confuse the general issue. Streams of unfamiliar people continued to recognize me. I began to get executive mail that would normally be for the vice president of a company. For instance, I would be asked to come to a conference concerning patents. I didn't know what they were talking about. I was also called to meetings with a certain executive. He always appeared very agitated whenever we spoke.

Most of the inquiries I received from these people were about the Moonbeam Project. I didn't know what it was. But one day, I had an intuitive urge. The basement of the BJM building in Melville had a very high security area. Consciously, I had no clearance to be in that area, but I walked in anyway. Normally, when you walk from one security area to another, you must hand the guard your badge and he gives you another badge (with a different designation). This permits you to walk in the secure area. I simply went in and gave him my badge from the lesser security area, and what do you know? He gave me a badge with my name on it! I'd had a hunch and it worked.

I walked around and let the churning of my gut determine what direction I should go in. I ended up in a posh mahogany paneled office. There was a large desk with a name plate on it that read, "Preston B. Nichols, Assistant Project Director". This was the first tangible physical proof I had that something out of the ordinary was definitely occurring. I sat at the desk and looked through all the papers. It was impossible to take the papers out of the place as I knew I would be searched very thoroughly on my way out of this high security area. So, I committed everything I saw to memory, to the best of my ability. I had an entire second career here that I knew almost nothing about! I can't even talk about most of it. It is top secret. I'm bound not to mention it for thirty years because of an agreement I signed when I went to work for BJM. However, I didn't sign a single thing regarding the activities of the Montauk Project.

Sifting through the material, I spent about six hours in my newly discovered office. Then, I decided I'd better get back to my regular job before the day was through. I handed back my badge and walked out. A couple of days went by before I decided it was time to go back and check things out again. Once more, I handed the guard my badge, but this time he didn't give me anything back. He said, "By the way, Mr. Roberts (fictitious name) wants to talk to you."

41

A man, Mr. Roberts, came out of an office that had "Project Director" written on it. He looked at me and said, "What do you want to come in here for, sir?"

"To get to my other desk," I replied.

He said, "You don't have any other desk here".

I pointed to the office where my desk had been. But as I entered the room with the Project Director, I found it to be gone. In the couple of days since I'd been there, they had removed every trace of myself from the room.

Somebody must have realized that I had visited my office when I wasn't supposed to. I had entered in an ordinary state of mind which was not to their liking. They apparently had not turned on the program (switching me to an alternate reality) for that particular day and must have been wondering why I'd shown up. They must have concluded that the process was leaking and that I was some how able to remember my alternate existence. As a result, they stopped everything. I was pulled aside through security channels and was told that if I breathed a word of what I thought I'd seen, I'd be locked up in jail and the key thrown away.

I tried to think of other strange incidents that had occurred. I'd kept a suspicious eye and had been investigating it for years. I was now sure that I had been experiencing two separate existences. How the hell had I been at Montauk and working at BJM, apparently during the same time period? I had already arrived at the conclusion that I must have been working two jobs simultaneously because there was a period of time when I'd come home and be totally exhausted.

At this point, all of what you've read was one huge confused mess in my mind. I knew that I'd been working on two separate time lines or maybe more. In fact, I'd discovered quite a bit, but it was more confused than clear.

42

I was, however, able to make a major breakthrough in 1990. I had begun constructing a Delta T* antenna on the roof of my laboratory. One day, I was sitting on the roof and soldering all the loops together into the relay boxes (which relay the signals from the antenna downstairs to the lab). Apparently, as I sat there and held the wires together to solder them, the time functions were causing my mind to shift. The more soldering I did, the more I became aware. Then, one day — bang! — the whole memory line blew open for me. All I could figure was that the Delta T antenna was storing up time flux waves as I was connecting it together. It just kept pushing my mind a little bit with regard to the time reference. The antenna was stressing time (bending it) and enough bend was created so that I was subconsciously in two time lines. This was my memory breakthrough.

Whatever the explanation, I was very pleased to have regained so much of my memory. I also believe my theory about the Delta T antenna is correct because the more time I spent working on the antenna, the more memories came back. By early June of 1990, all my key memories had come back.

In July, I was laid off. Subsequent to my firing, all of my close connections were removed as well. After having worked at BJM for the better part of two decades, I no longer had any links to or friends in the company. My information sources had been effectively severed.

You now have a general idea of the circumstances whereby I regained my memory. The next part of the book will contain the history of the Montauk Project that includes a general description of the technology involved.

---

* A Delta T antenna is an octahedronal antenna structure that can shift time zones. It is designed to bend time. Delta T = Delta Time. Delta is used in science to show change and "Delta T" would refer to a change in time. More about the nature of this antenna will be covered later in the book.

It is based upon my own memories and the information that has been shared with me by my various colleagues involved with the Montauk Project.

Editor's Note: (from Peter Moon)

When I met Preston Nichols, it was shortly after a huge layoff in the defense industry; and his job with AIL had been terminated. Around Long Island Psychotronics, it was no secret that he had worked for AIL. When it came to the book, however, he wanted no mention of it. There were several reasons for this, at least one of which is that he did not want to anger his former employers or create a security situation for himself.

Personally, I had never heard of AIL before I met Preston. For all I knew, it was a company he made up in his head. I was soon, however, to be disabused of this idea and only by rather strange reasons that naturally occurred in my own life.

Not long after signing a formal agreement to write "The Montauk Project" with Preston, I received a phone call to do some design work from Magnavox. Upon going to their building, I was shocked to see that it was adjacent to a rather huge campus for Airborne Instruments Laboratory or AIL. Although it is not such a big deal, it was as if the company itself was being imprinted upon my consciousness as a portent of things to come.

Not too long afterwards, I decided to join a writing club to sharpen up my writing skills and was directed to a lady named Charlene. She invited me to her house one evening where there would be gathering of writers, all of whom were aspiring to some sort of literary success.

To my surprise, bordering on alarm, there was a huge radio antenna on the house. As I was the first one at the gathering, Charlene directed me to her living room to wait where I promptly noticed a certificate that prominently featured "AIL".

"What was this?" I wondered.

Charlene explained that her husband worked for AIL. As she was also a little familiar with what I was working on, she suggested I wait until her husband got home which would be about 11:00 o'clock. His name is Dick Knadle, and to my surprise, he knew Preston Nichols pretty well. In fact, he had worked with him, heard his stories about Montauk and even visited Camp Hero in an attempt to investigate Preston's claims. The large antenna was a HAM radio antenna. Dick, however, was not a supporter of Preston's claims. His main argument, however, was that Preston was not an engineer but simply a technician. How could he possibly be qualified to participate in such a project?

Preston soon explained to me that he actually did secure an engineering diploma from the University of Tampa, but it was given on the basis of a series of "challenge exams" where you established that you had the equivalent knowledge thereof. Preston had gone to school previously at Brooklyn Polytech where he quit after being disgusted with his professor for stealing his work. Preston was and is a genius and has always been far ahead of his peers when it comes to technical matters. He said that he never let Human Resources know about his engineering diploma as it was more advantageous for him to work as a tech due to the fact that he would receive generous overtime pay.

Although Dick could not get on board with Preston's strange theories about the Montauk Project, he did mention one rather remarkable oddity concerning Preston. He said that although AIL typically had a very strict *dress code, it did not apply at all to Preston. In fact, he said it was as if Preston had no regard for the dress code whatsoever. This clearly indicates that Preston was considered extremely valuable. If he was a mundane technician, he would have been summarily fired for such.*

In this chapter, Preston also mentions a high security room at AIL which appeared to have strange coil structures and was part of "Project Moonbeam". At a book signing one evening, a man showed up who worked as a security guard at AIL. When he was told about this particular area and read about it in the book, he said he knew the exact location. Although he asked about it, he found out that it was indeed a high security area, and he knew better than to pursue it further.

Although the project in this chapter is referred to as "Project Moonbeam," that was not the original name. It was actually "Project Moonstruck." Preston was adamant about not using the actual name in the book. When I asked him why, he said that he did not want to piss off Dick Knadle. He said that Dick was over him in the organizational structure of the Montauk Project and that he led a secret life. Based upon my own conversations with Dick, I cannot verify Preston's claims. It is extremely coincidental, however, how all of these corroborative bits of experience landed in my lap, particularly when I was not looking for them. There was another bizarre coincidence which was perhaps even more to the point.

When I actually had completed the manuscript for this book and had the typesetting done, I looked for a local printer who could turn the book around in short order. When I finally found one that was willing to work on a fast turnaround, I visited the plant to meet the sales lady who promptly took me into the press room where there were huge stacks of promotional material, all of which prominently displayed "AIL", one of their primary clients. This was very weird. It demonstrated, however, that what I was dealing with was very close to home. Keep in mind, Long Island is 120 miles long with millions of people and numerous printers and defense contractors. The pattern of coincidences being played out was more than odd. I eventually had the book printed in the Midwest as the prices were much more reasonable, even when one factored in the shipping.

It is now over twenty-five years since "The Montauk Project: Experiments in Time" hit the bookshelves. It is hard to convey how challenging it was to compile the information and publish the actual book. There was considerable mental opposition from others with regard to disseminating Preston's story which was, after all, a controversial account of what he could muster up from repressed memories. It was like working in the shadow of Goliath or a great beast that everyone was denying the existence of but were also afraid to admit that it might not only exist but was influencing their lives.

T H Y R A T R O N

One of four pulse thyratrons that were used.
These drove the output tube. By supplying the pulse through the pulse
transformer to the output tube, the thyratrons regulated the frequency
hopping source. It was frequency hopping that made mind control and
bending of time possible.

7 WILHELM REICH AND THE PHOENIX PROJECT

The U.S. Government began a weather control project in the late 1940s under the code name "Phoenix". The information and technology for this came from Dr. Wilhelm Reich, an Austrian scientist who had studied with Freud and Carl Jung.

Reich was an extremely brilliant man but highly controversial. Although he experimented extensively and wrote many volumes, few of his critics have taken an honest look at all of his research because much of it is not available. Part of this can be attributed to the Food and Drug Administration who supervised a massive book burning of all his available materials and also destroyed much of his laboratory equipment.

Reich was known in part for his discovery of "orgone" energy which is orgasmic or life energy. His experiments revealed orgone energy to be distinctly different from ordinary electromagnetic energy. He was able to prove the existence of this energy in the laboratory. His findings were written up in various psychiatric and medical journals of the period. The discovery of a type of energy called "orgone" was not so controversial. It became very controversial with the powers that be when he reported curing cancer with his theories. He also associated "orgone" energy with "cosmic energy" and the Newtonian concept of "the ether". None of these views won him support from conventional scientists of the 1940s.

At the turn of the century, scientists had embraced the Newtonian "ether". This referred to a hypothetical invisible substance that was postulated to pervade all space and serve as a medium for light and radiant energy. Einstein, who embraced the theory in his early years, eventually determined that there could not be a calm ether sea through which matter moves. Not all physicists bought Einstein's argument, but Reich didn't disagree. He pointed out that Einstein disproved the concept of a static ether. Reich considered the ether to be wave-like in nature and not static at all.

Conventional scientists have since recognized the existence of phenomena that are a cross between particles and waves. They are sometimes referred to as "wavicles". Common research has also shown that vacuum space contains complex properties that are dynamic in nature.

Although it is not my cause to take up the case of Reich, his concept of the ether has proven itself functional in my research. It does not matter whether we are actually referring to "wavicles" or even more esoteric phenomena when we talk about the ether. It is the word that Reich used, and it is easier for me to use in describing this for the general public. The reader is invited to read up on Reich as his work is vast and encompasses much more than can be covered in the scope of this book.

For instance, he found practical uses for his theories such as modifying the weather. He found that violent storms accumulate "dead orgone" which he termed "DOR". Dead orgone refers to the accumulation of "dead energy" or energy that is on a descending spiral. Orgone and DOR were found to be present not only in biological organisms but also in empty regions of the environment as well. An active and enthusiastic go-getter would be considered to have plenty of orgone energy whereas a complaining hypochondriac who wanted to die would have DOR energy.

For example, he found that the more DOR in the storm system, the more violent the storm. He experimented with many forms of DOR busting and came up with a simple electromagnetic method to reduce the violence of storms. In the late 1940s, Reich contacted the Government and told them he had developed technology that could take the violence out of storms. Despite what disinformation you may hear, the Government already knew what Reich could do and considered him a brilliant man. They asked for his prototypes and he was happy to oblige since he wasn't interested in the mechanical development, just the research.

At this point, the Government's technology team merged Reich's discoveries with their own weather monitors and produced what is known today as the "radiosonde".

The Government's contribution to the radiosonde dates back to the "airborne metrograph"* of the 1920s. This was a mechanical device that recorded temperature, humidity and pressure. It was sent up in a parachute balloon and recorded information on a paper tape. The balloon was designed to burst so that the parachute would bring the metrograph back to Earth. The public were encouraged to retrieve them for a $5 reward which was considerably more money in those days. This was how the Government obtained data on the weather. As these devices were returned via the mail, the time that elapsed before the recorded information could be read was much too long.

---

* The word "metrograph" is more clearly defined if you understand that "metro" signifies that it was a meteorological device and that "graph" means to write.

51

In the late 1930s, a new device was designed that was called a "radio metrograph". This was similar to the airborne metrograph except that it contained electrical sensors. These sensors were connected to a transmitter that would transmit to a receiver on the ground.

The radio metrograph was the state-of-the-art weather device when Wilhelm Reich contacted the Government in the late 1940s. He gave them a little balsa wood package that could be sent up in a balloon. According to witnesses, approaching thunderstorms actually split up and went around the test site on Long Island.

The Government combined the technology of the radio metrographs with Reich's DOR busting device and called it the "radiosonde". It was developed until consistent effects on the weather could be reproduced. By the 1950s, radiosondes were being sent into the air en masse at a rate of about 200 per day.

Since these radiosondes were sent up in balloons, they would not come down hard enough to self destruct upon impact. The public would find them, and it would be impossible to keep the actual units secret enough without arousing suspicion. They publicized the apparent purpose of recording weather data which uninformed examination would back up. The real purpose was not that obvious. If someone tuned into one of these packages, the signal would not appear unusual when normal radio equipment was used. So far so good! They showed the public a data receiving station, set up to receive the inaccurate and unusable data. A small production run of this receiving equipment was produced.

There were literally hundreds of these radiosondes in the air every day. With the radio range being limited to 100 miles, there should have been a "pile" of receivers known as radiosonde receptors and they should have been very common. As I am a surplus radio collector "nut", it is quite strange that I have never seen a radiosonde receptor or the equipment that should accompany one. It is very unusual to have a data transmitter (in this case, the radiosonde) with no receiver to pick it up. This indicates that the Government didn't use the receivers!

My next clue was to look at the specification sheet for the radiosonde tube which emphatically states that the life expectancy is only a few hours (see page 60). Despite this, I have had a tube on the air for over 2,000 hours, and at this time have built over twenty such units with only one failure. This is a good industrial failure rate but is a major red flag. My only explanation is that if some local amateur radio operator finds or buys a radiosonde on the surplus market, he will read the data, get misled and not bother building a circuit that will run for "only a few hours". He will use another tube.

It appears that the Government does not want the public to use these tubes and find something unusual and thus blow their secret. This is why misinformation in the spec sheet preserves the secret. In fact, they are not telling a lie because the battery pack was designed so that the tube would burn out after three hours or so. This is caused by back bombardment of the cathode which would cool slowly and then destruct.

By the time these radiosondes hit the ground, they were dead. This way the public, who were encouraged to return them, wouldn't be able to pick up live units. If there was no secrecy involved here, why would the Government design a battery to burn out a costly tube that would have to be replaced after a very short usage? More disinformation was accomplished by packing the sensors in sealed vials which implies that upon exposure to the air, the sensors are short lived. Because of these precautions, the secret was maintained for over forty years which is excellent security.

Upon further examination of the radiosonde and its circuitry, I discovered that the temperature and humidity registers in the radiosonde didn't work. Not any of them! The temperature sensor was useless for recording the temperature, but it did have a function.* It acted as a DOR antenna while the humidity sensor acted as an orgone antenna. If DOR was sensed by the antenna, the transmitter would be broadcast out of phase and bust up the DOR and take the violence out of a storm. Conversely, transmitting in phase would cause the DOR to build up.

The humidity sensor had the same effect with orgone energy. Transmitting in phase would build up the orgone energy and transmitting out of phase would reduce it.

The radiosonde also contained a pressure element that would act as a switch signal and would maintain either DOR or orgone. This was how they built up the orgone energy.

The transmitter consisted of two oscillators. One was a carrier oscillator which runs at 403 MHz. The other ran at 7 MHz and is a relaxation oscillator. This one would pulse on and off depending on what was encountered. Somehow, this monitored the etheric function of the radiosonde. I haven't discovered everything there is to know about the radiosonde, but I have done a scientific analysis of it which I've included in the appendix (see Appendix A) for those who are interested.

What I have told you about the radiosonde is hard evidence that can stand up to scrutiny. It establishes the credibility of my story that there was a secret project that

---

* For those technically oriented, the temperature sensor is essentially a thermistor; but instead of being carbon based, it contains noble metals and exotic elements. It is a very poor temperature sensor because as the temperature cycles it up and down, the resistance curve changes and it doesn't hold its calibration. The humidity sensor suffers from the same problem.

involved weather control. We can't say exactly whether the radiosondes were used just to bust up violent storms, but the possibility was also there to build them up. The Government abandoned the weather control aspect eventually. Changing the weather, if it were proven in court, could lead to many law suits.

What is more intriguing than the weather aspect is the entire prospect of orgone and DOR energy and what could be done with that. In theory, this means that the Government could have targeted communities, buildings or an entire populace and transmitted orgone or DOR energy. These type of activities have been reported in Russia for years. Not much press coverage has been given the U.S. effort in this regard, but there has been some activity. Whether it has been used harmfully or in war, I cannot answer, but the potential was there. Forty years of development could also have made this a very refined technological device.

Please refer to Appendix B for additional information on Wilhelm Reich.

FIXED-TUNED OSCILLATOR TRIODE

6562
6562/
5794A

UHF pencil-type tubes having integral resonators; used in radiosonde service at a frequency of 1680 Mc. May be used at ambient temperatures ranging from -55°C to +75. Fixed-Tuned Oscillator maximum plate dissipation, 3.6 watts. The

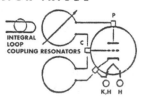

INTEGRAL LOOP COUPLING RESONATORS

==== *Technical Data* ====

6562 is a DISCONTINUED type listed for reference only. As a replacement, the 6562/5797A is directly interchangeable.

HEATER VOLTAGE RANGE°(AC?DC)........................ 5.2 TO 6.6 VOLTS
HEATER CURRENT (AT 6.0 volts)............................ 0.160 ampere
FREQUENCY (Approx.)....................................... 1680 Mc
FREQUENCY-ADJUSTMENT RANGE☐.................... ±12 Mc
This range of heater voltage is for radiosonde applications in which the ²ater is supplied from batteries and in which the equipment design requirements of minimum size, light weight, and high efficiency are the primary considerations even though the average life expectancy of the 6362/5794A in ³uch service is only a few hours.
☐As supplied, tubes are adjusted to 1680 ± 4 megacycles.

FIXED-TUNED OSCILLATOR

Maximum ratings:

DC PLATE VOLTAGE	120 max	volts
DC PLATE CURRENT	32 max	mu
DC GRID CURRENT	8 max	ma
PLATE INPUT	4 max	watts
PLATE DISSIPATION	3.6 max	watts
PEAK HEATER-CATHODE VOLTAGE	0 max	volts
AMBIENT-TEMPERATURE RANGE	-55 to +75	°C

Operating Frequency Drift:
Maximum Frequency Drift:
For heater-voltage range of 5.2 to 6.6 volts, plate-voltage range of 95 to 117 volts, and ambient-temperature range of +22° to -40°C............ +4 to -1 Mc

OPERATING CONSIDERATIONS

TYPE 6562/5794A may be operated in any position. OUTLINE 74, *Outlines* Section.

The flexible hater leads of the 6562/5794A are usually soldered to the circuit elements. Soldering of these connections should not be made closer than 3/4" from the end of the tube (excluding cathode tab). If this precaution is not followed, the heat of the soldering operation may crack the glass seals of the leads and damage the tube. Under no circumstances should any of the electrodes be soldered to the circuit elements. Connections to the electrodes should be made by spring contact only.

The 6562/5794A should be supported by a suitable clamp around the metal shell either above or below the frequency-adjustment screw. It is essential, however, that the pressure exerted on the shell by the clamp be held to a minimum because excessive pressure can distort the resonators and result in a change of frequency.

The plate connection should have a flexible lead which will accommodate variations in the relative position of the plate terminal in individual tubes.

The 6562/5794A may be mechanically tuned by adjustment of the frequency-adjustment screw located on the metal shell of the tube. A clockwise rotation of the frequency-adjustment screw will decrease the frequency, while a counterclockwise rotation will increase the frequency. The range of adjustment provided by the screw is ± 12 megacycles.

8 "THE PHOENIX PROJECT" ABSORBS "PROJECT RAINBOW"

While the Phoenix Project was investigating the weather and the use of radiosondes, Project Rainbow resurfaced in the late 1940s. Project Rainbow (which was the code name for the operation that brought about the Philadelphia Experiment) was going to continue research into the phenomena encountered on the *USS Eldridge*. This project was concerned with the "electromagnetic bottle" technology which eventually resulted in today's stealth fighter craft.

At about the same time, Dr. John von Neumann and his research team were called back. They had worked on the original Rainbow Project and went to work on a new endeavor. This was similar to the Rainbow Project but had a different goal. They were to find out what went wrong with the "human factor" of the experiment and why it failed so miserably.

In the early 1950s, it was decided that the remnants of Project Rainbow and the radiosonde project should be included under the same umbrella with the human factor study. After that point, the title of "Phoenix Project" was used to refer to all of these activities. The project headquarters was at Brookhaven Labs on Long Island and the first order of business was to put Dr. von Neumann in charge of the entire project.

Dr. von Neumann was a mathematician who came to the United States from Germany. He also became a theoretical physicist and was noted for his very advanced concepts of space and time. He originated the computer and built the first vacuum tube computer at Princeton University where he also served as the head of the Institute for Advanced Study.

Dr. von Neumann had what could be described as a "good technical feel". He had the ability to apply advanced theories to technology. His background in math gave him enough theory to communicate with Einstein, and he could in turn pass this on to the engineers and serve as a bridge between the two.

As von Neumann began work on the Phoenix Project, he quickly learned that he was going to have to study metaphysics. He had to understand the metaphysical side of man. The Rainbow technology had dissolved the physical and biological structure of human beings. People were stuck in bulkheads and changed beyond recognition in some cases. But it was the esoteric workings of the mind that had been affected first, in each case.

Von Neumann and his team spent about ten years working out why human beings had troubles with electromagnetic fields that shifted them through different places and times. They actually found out that humans are born with what is known as a "time reference" point. At conception, an energy being is attached to a time line and we all start from that point. To understand this, it is necessary to view the "energy being" or soul as distinct from the physical body of the person concerned.

Our whole reference as a physical and metaphysical being stems from that time reference which actually resides within the electromagnetic background of our planet. This time reference is the basic orientation point you have to the universe and the way it operates. You can imagine how you would feel if the clock suddenly started moving backwards and time as well. It is this time reference point that was thrown out of kilter with the individual crewmen of the *USS Eldridge* and caused them untold trauma.

The Rainbow technology turns on and creates what can be called an alternate or artificial reality. It creates a stealth effect by not only isolating the ship, but the individual beings as well, within a "bottle effect". Those beings were literally removed from space and our universe as we know it. This accounts for the invisibility of the ship and of the people on board. The alternate reality thus created has no time references at all because it is not part of the normal time stream. It is entirely out of time. To be in an artificial reality would be like waking up and not knowing where the hell you are. All of this would be very confusing.

The Phoenix Project was faced with solving the problem of bringing human beings into the "bottle" (and eventually out again) while at the same time connecting them to their real time reference (that they would know as the planet Earth, etc.). This meant that when they were in the alternate reality or "bottle", they had to be supplied with something that would give them a time reference. They solved this by feeding into the "bottle" all the natural backgrounds of the Earth — at least enough to convince them of a continuous stream time reference. To do otherwise would likely cause those in the "bottle" to experience transdimensional disorder and problems of this sort. This is why it was necessary to set a phony stage. They could then feel some degree of normality.

Dr. von Neumann was the ideal candidate for the job since he knew computers. A computer had to be used if they were going to calculate the time references of specific people and replicate those references while they were passing through an "electromagnetic bottle" or alternate reality. The people inside the "bottle" would be going through zero time and essentially a "no reality" or a disoriented one at best. The computer had to generate an electromagnetic background (or phony stage) that the physical being would synchronize with as well. If that wasn't done, the spirit and the physical body would go out of synch, thus resulting in insanity.

There are two points to be brought out here: the physical being and the spiritual being. This is why the time reference would lock in the spirit and the electromagnetic background would lock in the body. This whole project started in 1948 and was finally developed in 1967.

When this project was complete, a final report was written and submitted to Congress. Congress had funded this particular project thus far and followed the results. They were told that the consciousness of man could definitely be affected by electromagnetics; and additionally, that it would be possible to develop equipment that could literally change the way a person thinks.

Not surprisingly, Congress said no. They were concerned that if the wrong people got a hold of this technology that they themselves could lose their minds and be controlled. It is a very valid concern and word was given by 1969 to disband the entire project.

9 THE MONTAUK PROJECT BEGINS

It is no secret that Congress has tried to brow beat the CIA into finding out everything that goes on in the intelligence community. They have cut their funding, limited their legal powers, and even the most naive person would likely admit to a credibility gap of some degree. However, we are not dealing with the CIA proper here. Indeed, if the CIA is involved, it would be a splinter wing or wings that are being used by a source other than the CIA director.

When Congress disbanded the Phoenix Project, the group at Brookhaven had already built an entire kingdom around this project. They had Reichian and stealth technologies which could definitely affect the mind of man.

The Brookhaven group went to the military and informed them about this fantastic new piece of technology they were working on. They told them about a device that could make the enemy surrender without a battle simply by throwing a switch. Of course, the military was very interested. This was every war expert's dream. Imagine, a device that makes the enemy give up before the battle starts! *

The military became enthusiastic and were ready to cooperate. They were informed that they didn't need to get involved in the financing because that was covered by the group at Brookhaven National Labs. But, the Brookhaven people needed a place where proper experimentation could be done in seclusion. They needed certain equipment and

---

* I have included in Appendix C some evidence that suggests mind control devices were used against the Iraqis during the Persian Gulf War.

personnel from the military. They gave the military a list of all technology required.

Of particular import on the technology list was the old SAGE Radar. For this, they required a huge radiosonde that would operate around 425 to 450 Megahertz. From earlier research, it was known that this was one of the "window frequencies" for getting into the human consciousness. A very high powered radar device was needed that ran at 425 to 450 MHz.

The military had just what they were looking for: a mothballed Air Force base at Montauk Point that housed an obsolete SAGE Radar system that fit the bill. This system already had the RF sections and the modulator that would be required to build a huge radiosonde.

The SAGE Radar at Montauk was originally part of the early warning defense system used during the '50s and '60s. Today, satellites and over-the-horizon radar make this technology obsolete for defense purposes. It certainly raises an important question, even if one doesn't believe this story. Why was an old antiquated defense system turned on and utilized for a period of over ten years?

The name for this project was known as "Phoenix II" by the officials concerned. It has since been colloquially named by myself and others involved as the Montauk Project.

Up until then, Congress had been informed about what had occurred; but at this point, independent people were carrying forward with a project denied by Congress and were operating outside of any controls. They were even using the U.S. military in the process. Of course, it quickly becomes: "Who is using who?"

But, the point being stressed here is that it was being done without the supervision of elected officials and in spite of their objections.

The Montauk Base was being reopened. The SAGE Radar had been shut down since 1969/1970 when the base was turned over to the General Services Administration. It was a surplus government base without anything on it, and government financing for it had ceased.

It is obvious that major funding would be required for such an endeavor. The financing is shrouded in mystery, but it appeared to be totally private. I do not have documented evidence myself of the financing but have been told by my Montauk acquaintances that the original money came courtesy of the Nazis.

In 1944, an American troop train went through a French tunnel carrying ten billion dollars worth of Nazi gold. This train was dynamited in the tunnel while carrying 51 GIs. General George Patton was in Europe at the time and investigated this, but he couldn't understand how an American troop train could be dynamited in western Allied territory. As a general and human being, he cared about the GIs. The ten billion dollars was also a mystery, but Patton's efforts were blocked.

I've been told this gold eventually showed up at Montauk, and it was ten billion dollars of gold, then priced at $20 an ounce. This was the equivalent of almost 200 billion dollars in today's currency. It was used to finance the project initially and for years to come. After it was all spent, the project was allegedly financed by the infamous Krupp** family, who controlled the ITT corporation.

In late 1970 and 1971, the Montauk Air Force Base, 0773rd Radar Battalion, was actively being reestablished. They had to establish a staff, get the equipment working and set up the whole research facility. This took about a year, and by late '71, the Montauk Project was underway.

** The Krupp's were the owners of the German munitions factories for World War I and II. After being found guilty of war crimes and complicity with Hitler at the Nuremberg Trials, the head of the Krupp family was paroled from a light prison sentence and allowed to continue his notorious arms dealings.

The strictest security measures were employed, part of which were entirely valid. Although confidential stealth technology was involved, it is no secret that the stealth aircraft was designed with a radar resistant absorbine coating and a reduced surface cross section. What is secret are certain aspects of the "electromagnetic bottle" technology and how that was propagated. We're not going to discuss this or describe it as it remains a duly authorized military secret that concerns the defense of the United States. With this book, we are concerned with disclosing a project that should never have been activated in the first place. With no military or defense purposes to begin with, it was only designed for controlling the minds of the population and in spite of Congress forbidding this project.

The staff was a mixture of military employees, government employees and personnel supplied by various corporations. I was one of the latter and came to the project in 1973.

There were a number of Air Force technicians who had worked on the SAGE Radar in the '60s. The Air Force had assigned them to Montauk even though it was listed on the books as a decommissioned, derelict base. The technicians told the Phoenix group that they could change the general mood of the base by changing the frequency and pulse duration of the radar. They had noticed this as a professional curiosity after years of working with radar.

This was a surprise to the Phoenix people, and they found it very interesting. By changing the pulse rate and pulse width, they could change the general way people were thinking. This was what they were looking for.

This new information prompted what I now refer to as the "Microwave Oven" experiments. They took the reflector (which looks like a huge banana peel and which can be seen from a distance when you are at Montauk

Point), rotated it almost due west and angled it down so that it was focused on one of the buildings, in what they thought to be a safe place.

Inside that building, they had a chair inside a shielded room. First, they'd sit someone in the chair – this was usually Duncan Cameron. Then, they would open and close the door to determine how much UHF/microwave energy was getting into the room. All this was being done while the antenna was rotated and focused to a point in front of the building. At the same time, the transmitter was blasting gigawatts of power.

They experimented by running the transmitter at different pulse widths, different pulse rates, and different frequencies. They tried everything they could think of, just plain empirical experimentation. They just wanted to see what would happen to the person in the chair if he was bombarded by "x" frequency, pulse, etc. They observed that certain changes made a person sleep, cry, laugh, be agitated and so on. There were rumors that whenever the SAGE Radar ran, the mood of the whole base would change. This was very interesting to the project supervisors as they were primarily concerned with the study of human factors.

They wanted to see how they could train and change brain waves. This was done by changing the repetition rates of the pulse and the amplitude in correspondence to different biological functions. In this way, a person's thoughts could be controlled. With the 425-450 MHz of radio frequency power, they actually had a window into the human mind. The next step would be to find out what was inside of it.

Although the door to the shielded room was closed most of the time, it didn't work properly. The subjects were exposed to a strong enough field to influence the brain waves but not enough to do damage. However, if exposed to it for several days on end, it could be quite damaging.

Duncan sustained serious brain and tissue damage as a result of continuous exposure to 100 kilowatts of RF power at a distance of about 100 yards. The radio waves baked his brains and chest. Anywhere in his body where there was a change of density, zones of heat or energy would be created by the concentration of the microwave beams.

Upon visiting a doctor in 1988, Duncan's doctor commented upon the unusual scar tissue in his lungs. He'd never seen anything quite like it. Another doctor who was consulted said he'd only seen it in the service when someone had gotten in front of a high powered radar beam.

Previous research in or about 1986 indicated that Duncan was actually brain dead. Initially, I had asked different psychics to do readings on Duncan. They determined he was brain dead. I also knew that it was possible to inject a particular dye into the brain and have x-rays or CAT scans reveal what areas of the brain are using oxygen. Brain dead individuals suffer from a lack of oxygen to the cerebrum. If the psychic readings were accurate, his brain would not be using much oxygen.

I asked a neurologist with whom I was friendly with, and he said it was definitely possible that some one could be brain dead and yet be walking around. He cited some postmortems done on people in England and in the U.S. whose brains had unusual coatings inside of the skull. The coatings were about a millimeter thick.

More interesting yet was a case he'd encountered about ten years ago. He took out a group of x-rays of a normal human being and showed me the red areas. He also indicated blue areas but told me they were areas that didn't require much oxygen. Then, he put up another x-ray where the entire brain was blue. This meant that the person was alive and was walking around like a normal human except he has memory loss problems from it. He was essentially brain dead and the brain was using just enough oxygen to keep it from rotting. I noticed the corner of the x-ray and was surprised to see Duncan's name. Based upon this information, Duncan is indeed brain dead.

I asked the doctor for an explanation, but he wasn't sure. He could only offer a theoretical conclusion based upon psychic powers. He said that his profession recognized the existence of psychic phenomena but did not understand it.

At this point, we learned that the only reason Duncan is alive today is due to his strong psychic aptitude. The psychic part of his mind takes over the physical part of his mind and runs the body. His brain stem is alive; his spinal chord is alive; his body is alive, but his actual higher brain is dead. His psychic energy runs the body through the brain stem.

Duncan was not the only person affected. We don't know how many people were involved but the body count was probably high.

It wasn't until 1972 or '73 when it was finally realized that stealth technology dealt with non-burning radiation. One theory was that actual non-burning radiation, which is the higher order of components (as opposed to burning radiation), actually went through the reflector and would be opposite to the focal point of the antenna.

They tried it and turned the antenna around 180 degrees. They aimed the burning rays into the sky and hit the person with the non-burning rays. Then, they found they had the same mood altering capabilities, if not more than they had before, but this did not damage the people. But at what cost to the persons previously experimented on!

At this point in the project they were interested in monitoring people and changing their thoughts and moods, etc. It was not necessarily how they changed but the fact that they changed under certain circumstances. Different army units were invited to come to the base and have R&R (Rest and Relaxation) there. As far as the soldiers were concerned, it was free R&R in a beautiful location.

The outer base had a nice gymnasium and a bowling alley with excellent food and accommodations. Unbeknownst to the servicemen, they became guinea pigs for the mood control experiments. However, these were not the only guinea pigs. Experimentation was also done on the townspeople, including Long Island, New Jersey, upstate New York and Connecticut civilians, just to see how far it could go. Most experimentation, however, was done on the vacationing soldiers.

Time was spent monitoring different pulse types, trying this and trying that. They would note and categorize the different effects. It was all pure empirical experimentation and a huge data base was collected. Once they had enough data, they began to make some sense out of which functions did what.

During this period, they also experimented with frequency hopping. Frequency hopping consists of the transmitter instantaneously and randomly shifting around to any of five different frequencies (that were being fed to the transmitter). This point became very important later on as it was key to bending time.

They discovered that very fast frequency hops made the modulations more psycho-active. A data base was then developed that would list the frequency hop times (times you go from one frequency to another), how they pulse modulated, the rate the pulse modulated at, the pulse width, and the power output they pulsed it at. This was then coupled with the responding effects it had. The data base was very extensive and covered an extremely broad range of causes and effects.

After the extensive experimentation, they developed a control panel with which they could set different pulse modulations and timings. They knew that these different pulses and functions represented certain thought patterns from the individual. They could set the modulators and timings so that a transmission would be generated that would place thought patterns into an individual. This meant they could literally set this pulse at anything they wanted and expect a desired effect to take place.

All of this took about three or four years to research. The transmitter was now fully operational and hooked up. Programs could be typed in that would put the transmitter through its phases. Programs were derived that could change the moods of people, increase the crime rate, or make people agitated. Even animals within the vicinity were programmed to do strange things.

The researchers were able to derive programs whereby they could focus on a car and stop all the electric functions in it. I don't know what the modulations were, but I understand they found this quite by accident.

One day, there were military vehicles riding around the base. They suddenly ceased to operate without any function. An investigation ensued to find out what was occurring with the transmitter at the time, and a program was developed. At first, the program could only get the lights in a car to dim. It was eventually refined to where the program caused all electrical functions in a vehicle to cease.

Several years of research and collecting information had finally yielded a mind control device. The next objective was to create a precision technology with the material. In order to do this, help from very strange sources was enlisted.

Editor's Notes: (from Peter Moon)

While the actual financing of the Montauk Project will likely be shrouded in mystery for many more decades, if not forever, the theme of Nazi resources financing Montauk surfaced again when further research revealed that the notorious scandal surrounding Italian Freemasonry's infamous P2 Lodge started when Long Island's Franklin National Bank collapsed. This bank had been purchased by Michele Sindona so that he could launder money with regard to his ties to the Mafia and the Vatican. Perhaps some or a lot of this money ended up at Montauk. Later investigations revealed ties between the mob and Montauk. While this thread is clearly speculative, it reveals a financial swamp with infinite potentials.

Sindona, by the way, was a Sicilian educated by the Jesuits who was skilled in mathematics and economics but opted for a career in smuggling before being chosen by the Gambinos to launder their heroin profits. Giovanni Montini, who later became Pope Paul VI, was close friends with Sindona before the latter bought up banks with the Gambino heroin money. When Franklin collapsed, the investigation revealed that Sindona was a member of P2; and as a result, P2 lost its charter and was declared to be illegal by the Italian government.

On a separate thread, Preston's accounts of radio frequency being able to control or influence human emotion has been thoroughly corroborated and even documented. Martin Canon is one researcher who has addressed the subject although not necessarily with specific reference to the Montauk Project.

10 THE MONTAUK CHAIR

In the 1950s, ITT developed sensor technology that could literally display what a person was thinking. It was essentially a mind-reading machine. It operated on the principle of picking up the electromagnetic functions of human beings and translating those into an understandable form. It consisted of a chair in which a person would sit. Coils, which served as sensors, were placed around the chair. There were also three receivers, six channels and a Cray 1 computer which would display what was on a person's mind – digitally or on a screen.

It is still a mystery how this technology was developed. It has been suggested that the research was aided by the Sirians, an alien race who come from the star system known as Sirius. This theory has the aliens providing the basic design and humans working it out from that.

Three sets of coils were set up in a pyramid around the chair. There was also a coil around the top of the pyramid to parallel the base coil. The person would be placed inside the field of the coils. The three sets of coils were connected to three different radio receivers (Hammerland Super ProP 600's) and six outputs. An independent sideband detector, which had a floating carrier reference system, would provide six outputs from the three receivers. Three of them were of the sideband below the carrier wave. Three of them were of the sideband above the carrier wave. This brings to mind a very important question. If this device was reading minds, what was the carrier wave it was using to do this?

With the use of an oscillator, the detectors in the receivers were able to lock on a phantom or etheric signal that was being picked up by the coils. There was no actual carrier wave as we would normally know it. The detectors would lock in on the noise peak that the coils picked up from the three sets of frequencies the receivers were tuned to.

At this point, the research team was actually able to detect the signals that represented the comparable functions of the human mind. Solid signals that would change with a person's thoughts were actually coming out of the receivers. This device was actually reading the human aura which is a word that psychics and metaphysicians use to describe the electromagnetic field that surrounds the human body. In the same way that human speech is carried via radio waves, this device was carrying thoughts (which theoretically manifest in the aura).

The six output channels from the receivers were then run through a digital converter (turning them into computer language) and fed into a computer. A Cray 1 computer was used to decode what the receivers were picking up. A lot of hard work and a lot more computer crunching got things to the point where the computer could print out a dialogue. This would be a running dialogue of the person thinking.

More work got it to where the person would visualize something and a picture would actually appear on the computer monitor. Improvements and refinements continued until a 3-D representation of the actual audio/visual aspect (of the person's thoughts) appeared on the computer monitor and could in turn be printed out.

When the people at Montauk heard about this mind reading device, they thought it was great. They wanted to turn this mind reading machine into a transmitter. This could possibly cut the risks to human beings undergoing invisibility or time experimentation. The theory was that a person in the chair would transmit an alternate reality to the crew (like in the Philadelphia Experiment). When the ship became invisible, the crew would then be in synchronization with the alternate reality and wouldn't become disoriented or mentally lost.

At this point, a chair was procured which we now refer to as the famous "Montauk Chair". It was hooked up to the coil set-up from ITT. The Cray 1 computer, which was used to decode the transmissions being generated from the person in the chair, was then interfaced with an IBM 360 computer. This was, in turn, interfaced with the Montauk transmitter.

The IBM 360 was needed to control the modulation of the transmitter so that the transmitter could frequency hop across the entire band.

It was about this time that I remember Al Bielek taking on a key role. Al is one of the authors of *The Philadelphia Experiment and Other UFO Conspiracies*. He has memories of being involved in the Rainbow Project as well. Originally, he was brought to the project to explain what was going on metaphysically with the use of the transmitter on human beings. He was chosen because he not only had an engineering background, but he was psychically sensitive and had an extensive knowledge of esoteric matters.

It now became Al's job to help interface the Cray 1 computer with the IBM 360. The Cray 1 was putting out tons of information. They didn't know what to do with it and needed someone with esoteric knowledge to figure it out. They had to convert what the Cray 1 was putting out so that it would synchronize with what the pulse modulation computer wanted. The IBM 360 served this function and was essentially used as a translator and storage bank for what the Cray 1 was outputting. Al got very heavily involved because he was part of the team that figured out what program to put on the IBM 360 that would translate the Cray 1 output to drive the transmitter.

The transmitter had a modulation computer which was digitally fed the typical 32 bit code that the 360 put out. The modulation computer and the transmitter were set. The IBM 360 would tell the modulation computer how to modulate the transmitter. Now we had a system where one could put in 32 bit words of data and the transmitter would give back something. And here the chair fed the receptors feeding the Cray 1 which would tell what the person was thinking. They had to take this and translate what was coming out of the Cray 1 and make it so that the IBM 360 could re-encode the thought form that was actually transmitting. It took about a year to successfully link up the computers.

I had joined the project at this time to work with the radio frequencies and transmitter. Although some linkage had been achieved with the computers, they were having huge problems with feedback from the transmitter to the chair. The solution to the feedback was to move the chair down the coast to the ITT center in Southampton, Long Island. A psychic would then sit in the chair in Southampton and relay via computer to the Montauk transmitter.

The psychic would think thoughts, and the Cray 1 would decode them. They'd be put on a 32 bit radio link and sent to Montauk where they would go into the IBM 360. The IBM computer would then broadcast it out the transmitter and could build a thought form out at Montauk of what the psychic was thinking in Southampton. The device was essentially a mind amplifier.

It took another year of research before they could get a readable signal (based upon what the psychic was thinking at Southampton) sent to Montauk and out the transmitter. This was their first objective: get some thought fidelity from the chair through the Montauk transmitter and out the antenna. Besides Duncan, there were a couple of additional psychics on site. They literally tuned up the computer programs. Finally, the thought forms became clear. The psychic could concentrate on something in Southampton and the transmitter at Montauk would transmit a very clear representation of what he was thinking.

That was the first point at which the Montauk transmitter was working with high thought fidelity.

In another year, I recall as early '75, they discovered another problem. If there was a glitch in the flow of time in our reality, everything fell apart. In other words, if the psychic in the chair projected a reality (in terms of time in this case) that was not consistent with our reality (i.e. the flow of time in our reality), it would cause the connection between Southampton and Montauk to break up. Any glitch in space-time between the two cities would cause the transmission of the thought form to cease.

To better understand a time glitch, imagine time as a continuous pulsation or flow. As the basic pulsation of time interacts and changes form with other flows or phenomena, we have motion as we know it, against the backdrop of time. When these core pulsations that make up time are shifted (due to a reality change or other phenomena); the direction, speed, or flow of time is changed. This is what is known as a time glitch. Theoretically, these occur every now and then, and since we are referenced in our reality, we really don't notice a time glitch. Deja vu phenomena could well be an example of a glitch in the fabric of time.

With the chair in Southampton, the mind control experiments with the transmitter were not always working. This was attributable to the time glitches. It was also known that if a large amount of power was fed into the transmitter during a time glitch, there could be disastrous effects.

It now became imperative to get the chair working at Montauk. They first put tremendous shielding around the chair so that the electromagnetic fields at Montauk would not affect it. That didn't work, so they tried putting the chair in an electromagnetic dead zone. They picked the best dead zone available, but this was not successful either.

They worked through mid '75 but continued to have difficulties until they consulted the original prototype that the chair was based upon (allegedly devised by the Sirians). This device was not identical to the one ITT had created. It had a different kind of coil set-up wherein the coils were connected to crystal type receivers. These were actual crystals and not ordinary electronic devices.

After review of the prototype, secret bids for a new chair were put out and RCA came up with the winning bid. Nikola Tesla* had designed receivers for RCA in the 30s. Tesla's work during this period was done under the name "N. Terbo" which referred to his mother's maiden name. These Tesla receivers had very special coil structures. They were normal type radio coils but were arranged in strange coupling patterns as set up and designed by Tesla.

The set up of the Montauk Chair was also enhanced by using Helmholtz coils. These were placed around the chair to serve as pick up coils. In ordinary electronics, Helmholtz coils consist of two sets of coils. They possess a unique property in that they can be phased to create a constant field (of energy) inside the coils. At Montauk, the researchers extrapolated upon the principles of Helmholtz coils. They used three sets of coils (X, Y and Z),

---

* Nikola Tesla was an electronic genius who was the first to discover and apply the principles of alternating current. With the financial backing of George Westinghouse, he revolutionized the way electricity was used across the world. See Appendix D for more information on Tesla.

and phased them so that while a constant energy could be maintained inside the coils, there was absolutely no effect on the outside.

The coil structure in the receivers designed by Tesla was ideal for the Montauk Project. Not only would the chair be in a coil structure but so would the receivers themselves. This would shield the energy field.

It should also be noted that the coil structures in the Tesla receivers are also known as Delta T or Delta Time coil structures. The property of shielding an energy field is part of what enabled a "bottle effect" to be created around the *USS Eldridge* in the Philadelphia Experiment. These Delta T coils were actually picking up three axes of time signals. More pertinent to the project, they no longer had a microwave link that would malfunction during a reality shift.

To get the Montauk chair operating without interference, they had to replicate what the crystal receivers did with the "Sirian" technology. The coil structures in the prototype receivers were Delta Time coil structures. And the receiver itself did the Delta Time function but not the antenna. ITT had the Delta Time function in the antenna instead of the receivers. The RCA version used standard type Helmholtz pick up coils that could accomplish Delta Time conversion in the receivers. They also had the same kind of detector system and oscillator locks that ITT used with the Cray 1 computer.

At this point, it now became inclusive of the coil only. Outside the coil structure, there was no sensitivity. They could put the chair in the dead spot that was between the transmitting antenna on top of the transmitter building and the transmitting magnetic antenna that was underground. This was in the underground basement of the transmitter building which had already been tightly shielded. In the next room, they had these three specially designed

77

receivers with another rack of equipment. These were used to synchronize all local oscillators with the signal, similar to the ITT system.

Now, the antenna, the transmitter, and the chair were in the same time plane. The computers were in their own time plane. It didn't matter that they had the chair underground and the Cray and 360 in the other building (feeding back to the transmitter building). When everything is digitized, one is no longer in real time. A "fake time" is created. The computers could have been located anywhere. The computer building was designed to operate computers and shielded out electromagnetics and energy from the antenna so that the electromagnetics and energy didn't drive the computer insane. The operation center was totally shielded in cement and steel.

Finally, they created the second and last generation of the Montauk chair. It performed the same purpose as the first chair. It brought the same six channels of information to the computer, but there was an additional advantage. It was immune to the signal from the antenna. Now, the signal from the antenna didn't feed back and cause interference. So, they had everything on site. They spent another six months until about late '75, early '76, just aligning, adjusting, and making sure everything was working.

They finally got the transmitter functioning which was quite astounding. What happened afterwards was even more so.

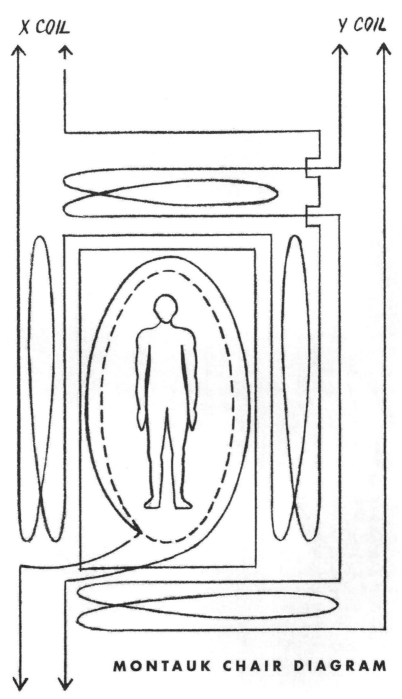

X COIL

Y COIL

MONTAUK CHAIR DIAGRAM

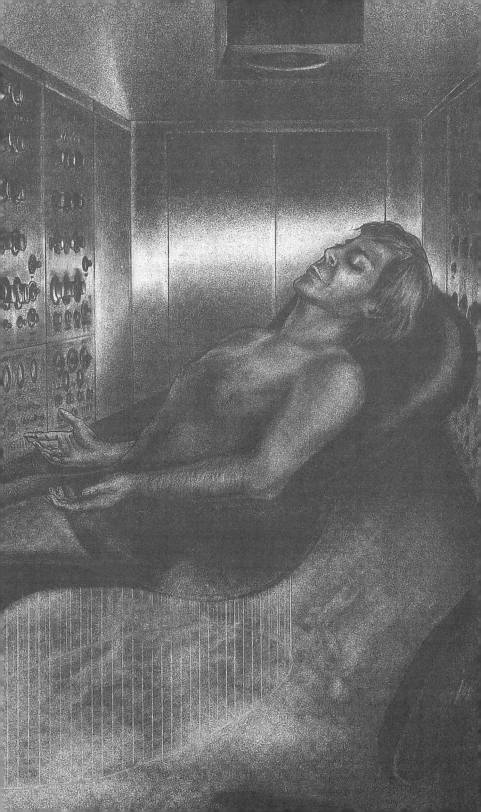

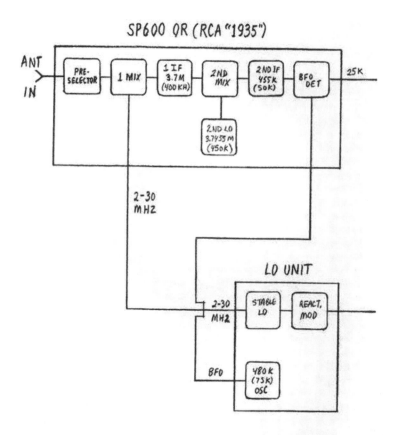

SP600 OR (RCA "1935")

MONTAUK CHAIR RECEIVER
VERSION 1 SP600   VERSION 2 (RCA)

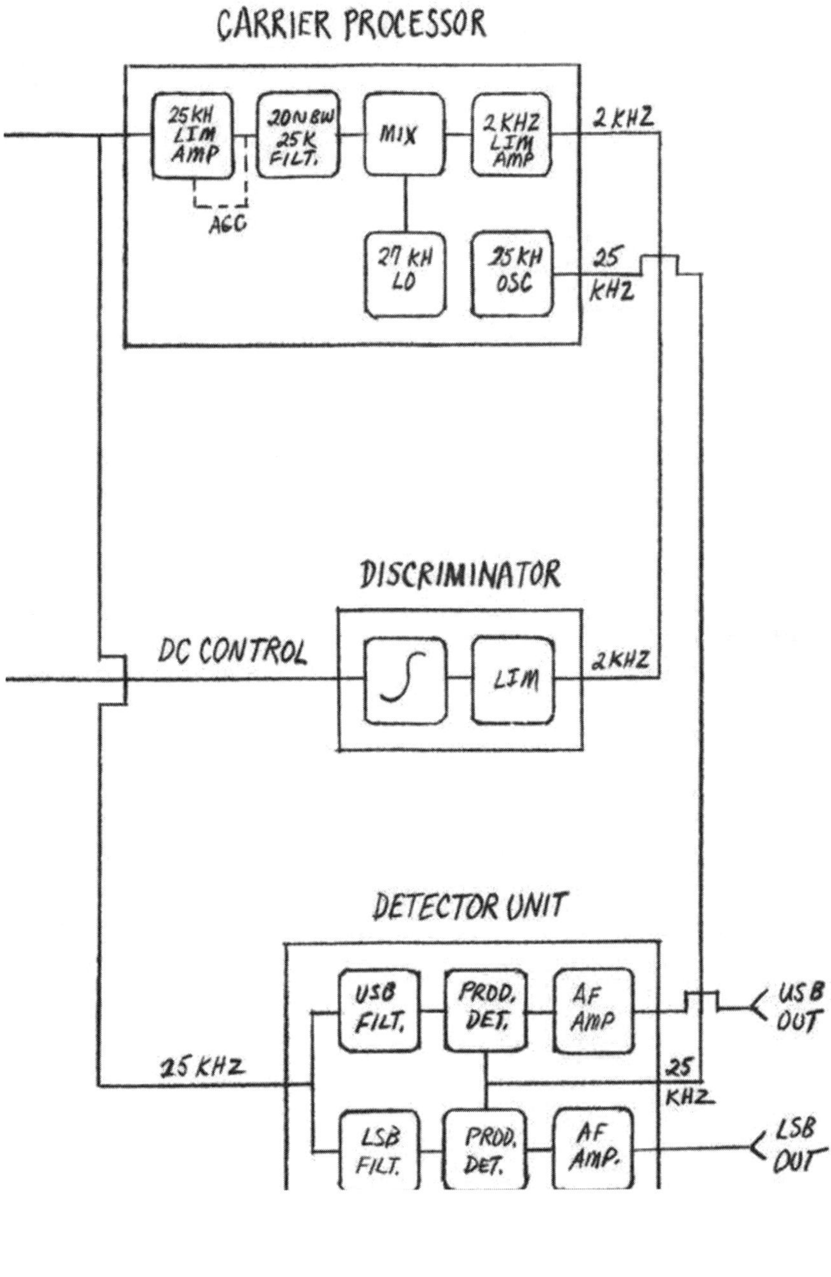

CARRIER PROCESSOR

| 25 KH LIM AMP | 20 N BW 25 K FILT. | MIX | 2 KHZ LIM AMP | 2 KHZ |
| AGC | | 27 KH LO | 25 KH OSC | 25 KHZ |

DISCRIMINATOR

DC CONTROL

2 KHZ

DETECTOR UNIT

| USB FILT. | PROD. DET. | AF AMP | USB OUT |
| LSB FILT. | PROD. DET. | AF AMP. | LSB OUT |

25 KHZ

25 KHZ

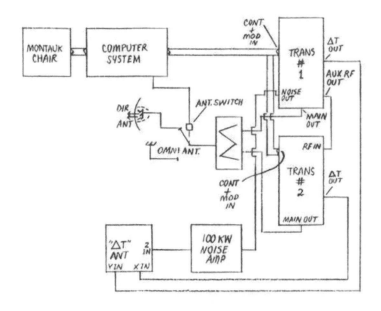

OVERALL BLOCK DIAGRAM

11 CREATION FROM THE ETHER

Once they had the transmitter working, it took about another year to work out the computer programs so the system would receive and transmit all psycho-active functions. By late '77, the transmitter was reproducing thought forms without glitches and with a very high degree of fidelity. At this point, they pulled out all the stops. They had the psychic, Duncan Cameron, concentrate on a solid object, and guess what happened? The solid object actually precipitated out of the ether!

In his mind, he would concentrate on a solid object, and it would appear somewhere on the base. Whatever Duncan would visualize, the transmitter would transmit the lattice (or matrix) for and build enough power to materialize whatever he was thinking of. Every single point to where he could witness to a particular spot on the base, at that spot an object would materialize. In other words, if he would hold an object in his hand and/or visualize it, it would appear at the given spot. They actually had discovered pure creation out of thought with the use of the transmitter.

Whatever Duncan could think up would appear. Many times, it would be only visible and not solid to the touch, like a ghost. Sometimes, it was a real solid object that was stable and would stay. Other times, it was a solid object that would remain as long as the transmitter was turned on and then fade out as the transmitter was turned off. The read out from the computer gave an accurate representation of what Duncan was thinking. The researchers could then select what thoughts would be broadcast out of the transmitter. Most of these thought forms were broadcast in the vicinity of the Montauk Air Force Base, but other locations were used as well.

What Duncan thought of as a subjective reality would be created as an objective reality (either solid or transparent, depending on the circumstances). For example, he could think of an entire building and that building would appear on the base. This type of experimentation was routine.

The system worked with a good degree of fidelity. Now, they wanted to see what they could do with it. The first experiment was called "The Seeing Eye". With a lock of a person's hair or other appropriate object in his hand, Duncan would concentrate on the person and be able to see as if he was seeing through their eyes, hearing through their ears, and feeling through their body. He could actually see through other people, anywhere on the planet. This style of experimentation was extensive, but I don't how far it was taken.

It is truly incredible that such a feat could be accomplished, however, the agenda employed was more sinister than incredible. They were interested in controlling how human beings think. The next move was to see if they could put thoughts in the head of another person. For instance, they would have Duncan meet a subject individual. Subsequent to the meeting and unbeknownst to the individual, Duncan would concentrate on the individual. Ninety-nine percent of the time, the subject would get thoughts similar to Duncan's. Being able to push his mind so far into the mind of another being, Duncan could control another person and make them do anything he wanted. This control factor was on a deeper level than ordinary hypnosis.

Through Duncan, the equipment and the Montauk transmitter, scientists could actually load information, programs and commands into an individual's mind. Duncan's thoughts would become an individual's own thoughts. And, using this process, an individual could be made to do something he wouldn't ordinarily do. This was the start of the mind control aspect of the Montauk Project.

This line of research continued until about 1979. Many other different experiments ensued. Some of them were interesting, but others had horrible consequences. They would target individuals or masses of people, animals, places, and technology. They could basically target anything they wanted. For example, a TV set could be made to go haywire. They could stop the picture or shut if off entirely. They telekinetically moved objects and destroyed rooms.

In one particular case, Duncan concentrated on shattering a window. Enough force was generated to the point where it actually broke a window in the nearby town of Montauk. Animals could also be made to charge off Montauk Point and into the town. Humans could be influenced to start a crime wave.

One must realize that when Duncan did these experiments, he was in an altered form of consciousness. He had been given special training which could possibly have been administered by the CIA or NSA. In any event, his conscious mind would be diverted through sexual bliss. What could be termed the primitive mind would then surface. Duncan, the individual, would be transferred into an orgasmic trance. His primitive mind, at the disposal of the researchers, became very suggestive and therefore controllable.

For this programming, information could be installed via any of the body's senses. Duncan would then be directed to have his primitive mind concentrate on the information thus installed. For instance, once his primitive mind surfaced and was told to concentrate on something, it would concentrate with its whole being. His whole mind would focus on one subject while his body went into suspended animation.

The primitive mind could also be cleared of previous programming, and something else could be inserted. There was a literal translator whereby they could program in whatever they wanted. Spoken words, written words, movies, music or whatever was needed was employed to work the primitive mind.

These techniques were the key to getting clear thought forms from the transmitter that would either affect another person's mind or bring creation out of the ether.

By 1978, the mind control techniques were fully developed and recorded. Appropriate tapes were made and distributed to different agencies so they could be developed into something practical.

12 TIME WARPING

As the experiments continued throughout 1979, a very peculiar phenomena was noticed. As Duncan's thoughts were projected out through the transmitter, they would suddenly cease. This was disappointing and appeared to be a malfunction. Eventually, it was noticed that the projection of Duncan's thoughts hadn't ceased. They were just occurring out of the normal time stream!

For example, he would concentrate on something at 8:00 PM and the object or occurrence would happen at midnight or even 6:00 AM. Whatever he thought of would not happen at the time he thought of it.

It now appeared that the Montauk scientists could now use Duncan's psychic powers to actually bend time!

They eagerly started to research this phenomena. We were all required to attend what were known as the "Sigma Conferences" which were held near Olympia, Washington. These conferences were on the subject of time functions, and we were there to gain a better knowledge of how time works. We were told we had to optimize the use of the transmitter for time manipulations.

We learned that the equipment being used was strong enough to bend time, but it wasn't doing a complete job. The antennas being employed were giving us what could be a side effect of "time warping". This side effect of time shifting did show, however, that the basic equipment was sufficient to do it. But, we required an antenna that was much more effective in creating time potentials.

After going to several conferences and talking to many people, our research group decided that the radio frequency being used was not working. Changes had to be made such as setting up pulses into a coil. We also studied pyramid based geometry and how to use that to bend the time field. Additionally, we had to learn more about what is known as the Delta Time function (time changing function).

The key clue to our understanding time was a suggestion that we use a particular type of antenna structure, which I now refer to as an Orion Delta T antenna. It is referred to as "Orion" because there was a persistent rumor that the design was given to the project by aliens from the Orion constellation (this is a different group of aliens from the Sirians whose knowledge was allegedly used for the Montauk chair). According to the rumor, the Orions knew we were close to achieving our task and had their own agenda for helping us.

The Orion Delta T was a huge octahedronal antenna, and it was placed underground. Its height was about 100 to 150 feet from point to point. Excavations were completed to about 300 feet to house the antenna under the transmitter.

The Montauk chair was placed under the transmitter and above the Delta T antenna. This was done in order to phase the above ground RF antenna with the below ground loop antenna so that the chair was in a null point between them. The null point was meant to cut out the interference even deeper. It knocked the interference right out of the chair – completely.

The Delta T transmitting antenna was supplied by three drives. Two of the drives came from the pulse modulators of the two transmitters and fed into the x and y coils of the Delta T. (The same pulse that supplied power to the amplitron also supplied power to the Delta T antenna that was underground). The third axis was the z axis. It was placed around the perimeter of the antenna and was derived from a white noise source that came from a 250 kilowatt audio amplifier. The white noise correlated the whole transmitter and more will be said on that later.* The RF was fed into an omnidirectional antenna located above ground on the top of the transmitter building. Additionally, the non-hertzian component (which is etheric in nature) of the RF made it below ground and interfaced with the magnetic field that had been generated underground. When these frequencies are summed in that manner, time disturbances and distortions result.

The basic techniques were the same as those employed in the Philadelphia Experiment. On the *Eldridge*, they had the RF transmissions on the main mast of the ship. The coils were placed around the deck and were driven by pulses. We had essentially duplicated but upgraded the Rainbow Project machine. This technique also made the project far more controllable.

In addition to the Delta T antenna, there are two other key points to understand: zero time and white noise.

Zero time was referred to previously, but I will give a more complete understanding of it now. First, zero time is outside the realm of our normal three dimensional universe. It would be considered senior to the created world as zero time existed prior to our created world. Zero time is our basic connection to the universe.

As our universe rotates, it rotates around zero time. But our universe is not the only one. Every universe has a zero point. All the zero points of the different universes coincide and never move: that is why it is called a zero point.

It may help to imagine a carnival style merry-go-round that revolves around a central booth. The man inside

* White noise is an impulse at every frequency at the same time. When you are tuning your FM radio dial, the noise you hear between stations is white noise. It can be thought of as a sudden burst at every frequency or a bunch of impulses thrown together.

that booth would represent the zero point. In addition to the merry-go-round, there would be several more merry-go-rounds at different levels, but all would be under the control of the central zero point booth.

A zero time reference generator had already been constructed by Nikola Tesla in the 1920s. It consisted of an assortment of spinning widgets and rotating wheels. Colloquially, we referred to it as a "whirligig". It is a strange device because when you turn it on, you can hear it "lock in" to something, but we are not referring to the power line. I'm told it locks into the rotation of the Earth itself which is a secondary zero time reference. It is secondary because the Earth's rotation is inertially related to the solar system which is inertially related to the galaxy, on down to the universe. The universe rotates around the zero time point.

One can get an even better understanding of this by reading up on Tesla and how he discovered alternating current by applying the principles of the rotating magnetic fields of the earth. The zero time generator is to some degree an extrapolation of that, however, it doesn't just refer to the rotation of the Earth. It takes into account the orbit of the Sun, our galaxy, and ultimately the center of our whole reality.

The other key point to understand is white noise. White noise could be considered the glue that makes the whole operation work. It basically made the whole transmitter system coherent. It is a highly technical operation which I will simplify.

The SAGE transmitter contained something like forty or fifty crystal controlled oscillators, mixers and amplifiers that generated a 425 MHz signal. It also had "frequency agility", which meant it was able to spontaneously switch from one frequency to another.

Along with the transmitter, they had what is called a "COHO" or a "coherent oscillator set-up". Normally, a

"COHO" would function by having only one frequency reference. However, this is not how the Montauk transmitter achieved coherency.

In order to make it entirely coherent, we took every oscillator available and amplitude modulated it with white noise. Since white noise is fifty per cent correlated to everything, it serves a universal auto-correlating function. The result was that all of the etheric components of the oscillators were now coherent to each other. We weren't trying to correlate the normal electrical functions as they didn't concern us. We were only interested in the etheric functions as they gave the results we were searching for.

A very stable time reference was required from the zero time generator. This produced two 30 hertz waves, referenced to zero time. One was connected to the computers and synchronized the clock or timing functions. The other modulated the white noise generator. By adjusting the phase between them, we could focus on and monitor the whole operation. This enabled us to take the correlations of the white noise and refer it right to the center point of time where all time crosses.

The purpose of this experiment was to make the psychic transmissions of Duncan time coherent. Dr. von Neumann had instructed us that the transmitter had to be time coherent with respect to zero time. The whirligig zero time reference also served as a witness point back to the Philadelphia Experiment, and that was very important. The project was trying to open a time door to the USS Eldridge in 1943.

Modifications continued on the equipment through 1979, until we had a coherent transmission system with respect to time phase.

Now, they had to calibrate Duncan. This meant they had to adjust and modify the equipment to synchronize with him. He had already demonstrated that he had zero point references of his own when the inadvertent time bending had occurred. This could perhaps be better explained due to his prior experience during the Philadelphia Experiment. There, he had jumped off the *Eldridge* and was thrust into a time vortex. At Montauk, he was now in an entirely new set of circumstances, but his familiarization with zero time had apparently never left him.

There were also other psychics, but Duncan was the first they had used, and he was in the chair ninety per cent of the time the system was in operation. If he was sick or didn't feel well, they'd wait a day. Because every time they changed the operator, they had to recalibrate and reprogram the computers and pulse modulator, and it took about two full days to do that. If Duncan was out for two weeks or more, they'd put in another operator, but I only remember one time when they did that. It was almost a disaster because they didn't spend enough time in the initial calibration. From then on, Duncan was the one and only one who ran the equipment. A backup had to be there, however, in case something happened to Duncan.

By 1980, the big radar reflector (that looks like a huge banana peel) on top of the building was no longer in use. Now, there were two transmitters that fed the omni-directional antenna (the one above ground). The pulse modulators of the transmitters were feeding both that antenna and the coils of the Delta T antenna (underground).

Also connected to the computer was the Montauk chair which was now placed between both antennas at the null point. By this time, the computer system was huge and was housed inside the control room next to the radar tower. Additionally, the computer room contained a lot of different terminals and displays to monitor the various activities of the project.

Duncan would start out sitting in the chair. Then, the transmitter would be turned on. His mind would be blank and clear. He would then be directed to concentrate on an opening in time from say, 1980 (then the current time) to 1990. At this point, a "hole" or time portal would appear right in the center of the Delta T antenna – you could walk through the portal from 1980 to 1990. There was an opening that you could look into. It looked like a circular corridor with a light at the other end. The time door would remain as long as Duncan would concentrate on 1990 and 1980.

I've been told by those who entered the tunnel that it looked like a spiral similar to science fiction style renditions of a vortex. When outside the tunnel, it looked like you were looking through space – from one circular opening through space to a circular but little bit smaller window at the other end. I was considered too valuable to the technical operation and was not allowed to travel through the portal.

From 1980 to late 1981, the time function was calibrated. At first, the time portals would drift away. One might go through the portal and come out in 1960. But when one went back to find it later, although it was still being tracked in real time, the portal would not appear where it should have been. One could easily get lost in time and space. Initially, the portal would be opened up, but it would drift. This was because Duncan himself was drifting. He had to go through extensive training to get the portal to be stable. We also had to focus the transmitter more closely and tighten up the thought form translation to get everything right. We would spend days just trying to get a particular time change to occur as predicted. However, there was no particular problem with creating a time warp. Predicting what it would do was the difficulty. Finally, towards the end of 1981, we learned how to stabilize it so that when a portal would appear, it would remain. Although the function was not absolutely perfect, it was predictable, stable and running according to plans.

Essentially, what the scientists were doing was using the 1943, 1963, 1983 vortex which was based upon the natural twenty year biorhythms of the Earth. 1943, 1963 and 1983 acted as anchor points for the main vortex. Sub vortices or open ended vortices would be created by going from the main one through an anchor point ('43, '63 or '83). At Montauk, August 12, 1983 was used.

For example, let's say they wanted to reach November 1981. There would be a bridge point from November of '81 to August 12, 1983. From August 12, '83 they could go to whatever time they wanted. The vortex ran between August 12, 1943 and August 12, 1983 because that was the master vortex. That gave them the stability to create what we call an open ended vortex. It is called open ended because there is no device at the other end which is anchoring it.

Although they had stabilized the time aspect of the portals, they had to work on the spatial aspect as well. They stabilized this aspect so that they could not only place a portal at a particular time but in a particular space.

Once time stabilized and the above was accomplished, they kicked out everybody and cleared the entire base except for a few key persons. I remained there as I was the technical operator and was essential to the project. Duncan remained as he was the psychic who made the operation work. The entire system was tuned to him. Two other psychics were also retained as back-ups in case Duncan was killed or incapacitated. The project directors also stayed, but the military left. A whole new team was brought in to do the more mundane functions of maintaining the base.

Up to that point, everyone operated on a "need to know" basis. Security was already tight, but they wanted even higher security. They didn't want the military to know what they were doing with time. But everybody knew there was something weird going on. They just didn't know what.

13 TIME TRAVEL

As most of the technicians were gone, a new technical crew was brought in. I don't know who they were and what their qualifications were, but they were called the "Secret Crew". The project was relaunched and is now sometimes referred to as "Phoenix III". This lasted from February 1981 until 1983.

The objective now was to explore time itself. The crew began to look at past history and to the future, just scouting around. They would search ahead for a hostile environment. Through the vortex, they could sample the air, the terrain and everything without entering the portal.

Those who travelled through the vortex described it as a peculiar spiral tunnel that was lit, all the way down. As one started to walk down, he would suddenly be pulled through it. It propelled one out the other end, usually in another place (as opposed to Montauk), or according to where the transmitter was set or placed. It could be anywhere in the Universe.

The tunnel resembled a corkscrew with an effect similar to lit bulbs. It was a fluted sort of structure and not a straight tunnel. It twisted and took turns until you'd come out the other end. There, you would meet somebody or do something. You would complete your mission and return. The tunnel would open for you, and you'd come back to where you came from. However, if they lost power during the operation, you'd be lost in time or abandoned somewhere in the vortex itself. When someone was lost, it was usually caused by a glitch in hyperspace.* And although many were lost, the scientists didn't abandon people deliberately or carelessly.

According to Duncan, there was also another function of the time tunnel. About two-thirds of the way down the tunnel, one's energy leaves the body. One would feel a big thump accompanied by a tendency to see on a broad scale. He reported sensing a higher intelligence along with an out-of-body experience. This was referred to as a FULL OUT. The researchers would try to manifest this in Duncan. It could have been for further "Seeing Eye" experiments or for other reasons.

It was routine to create a tunnel, grab somebody off the street and send them down. Most of the time these people were winos or derelicts whose absence wouldn't create a furor. If they returned, they would make a full report on what they had encountered. Most of the winos used for the experiments were sobered up for a week before entering a portal, but many didn't make it back. We don't know how many people are still floating around in time, whenever, wherever, and however.

As "Phoenix III" developed, the individuals so chosen for this research would be wired up with all sorts of TV and radio equipment so that they could report back "live". Each individual would be escorted through the portal, sometimes with force. TV and radio signals would travel right through the portals and as long as they could pick these up, researchers would have radio/video tapes of what the time traveller had experienced.

Those controlling the project began to play all sorts of games, manipulating the past and future. I don't really know what they did because I was the one at the switch. My station was in the transmitter building, and I had to keep everything going. I was not privy to a lot of what was going on, but at one point I do know that they had an extensive library of videotapes. I saw the tapes themselves

---

* Hyperspace is defined as space which exceeds the boundaries of three dimensions.

although I was not granted extensive viewing privileges. Actually, I designed and built the viewer (with the aid of tremendous resources) so I had some idea of what was going on. Much of what I knew came from Duncan's own reports because, by that time, we had become good friends. Eventually, we were debriefed and sent on our separate ways. Most of my memories of him had been wiped out.

In addition to the derelicts, the researchers also used kids for some reason. I'm not sure what exactly the purpose was, but there was one kid at Montauk who would go out and get other kids and bring them to the project. He was like a tractor beam. He lived in Montauk and would circulate around very effectively. There was also an entire corps of these around the New York metro area that could get away for six hours or so without being missed. They were specifically trained to go out and bring in other kids. Some kids returned home, some didn't. The kids chosen were between 10 and 16, or maybe 18 at the oldest and 9 at the youngest. Most were just about to reach puberty or had just finished it. They were usually blond, blue eyed, tall and light skinned. They fit the Aryan stereotype. To my knowledge, there were no girls in this group.

A later investigation showed that Montauk had a NeoNazi connection and that the Nazis were still on the Aryan kick. We don't know where the kids went, what they were educated in or programmed for. Whether they came back or not is still a mystery. What information is available is that they sent every raw recruit into the future to 6037 AD, always to the same point, to what appeared to be a dead city in ruins. Everything was stationary, not unlike a dream state. There were no signs of life. In the center of the city was a square with a gold horse on a pedestal. There were inscriptions on that pedestal, and recruits were sent there to read what they said. Each recruit would interpret and report. We still don't know what the researchers were after. They could have been trying to find the same answer from different people. I don't know. Duncan suggested there was technology in the pedestal and that they were trying to get somebody to sense or feel what the technology was.

Someone else involved in the project has said the horse was there to test the powers of observation of the recruits and that it also served as a point of reference. The recruits were always asked if they saw anybody in the city. Each individual would interpret what he observed and report.

We know a lot of people were shoved somewhere into the future, maybe 200 or 300 years ahead. Estimates range from three to ten thousand people that were eventually abandoned. We have no idea for what purpose.

I have already said that I don't know exactly what they did with time. I wasn't there, but I do know they did a lot with World War I and World War II. They monitored those times and took pictures. They knew exactly what they were doing. They could actually make up a secondary vortex to observe what was going on. We called this a seeing eye function. The original vortex was such that one could drive a truck through it. But the secondary vortex was an energetic vortex, with no physical solidity. One could, however, beam through it. Using phase conjugation through the elaborate computer set up, past and future history could actually be transmitted through the portal and viewed on television.

14 MISSION TO MARS

The project researchers continued to scout around in time. It was in late 1981 or '82 when the first actual use of this technology was employed to gain entrance into the underground areas in the big pyramid on the planet Mars.

As this material will be controversial to much of the general public, I will try to give some background.

There is currently a video tape in circulation entitled "Mars Mission". This is a presentation to NASA scientists by scientific journalist Richard Hoagland concerning the tetrahedral complex that is associated with the "Face on Mars". In this video, Hoagland shows the "face" and nearby pyramids that were photographed by the Viking spacecraft in the '70s. Computerized projection techniques are used that give one a 360 degree "fly by" of the "face". The video also gives a close look at the pyramids.

Hoagland is trying to convince NASA to make more pictures of this region which is known as Cydonia. NASA has been hard to convince and minimized the significance of Hoagland's work. In fact, a major effort was launched to prohibit the showing of this tape on public television stations. The story of this scandal was reported on by New York radio station WABC.

Why would NASA assume such a stance about a subject that is so intriguing?

The answer is perhaps explained by a book entitled *Alternative 3* by Leslie Watkins with David Ambrose and Christopher Miles. This book was based on a 1977 video that revealed a secret space program being run by an international conspiracy that included both the Russians and the United States. It is a fascinating account that includes astronauts breaking security, disappearing scientists, murder and the establishment of slave societies on the moon and the planet Mars. The book claims that men actually landed on Mars as early as 1962.*

It is not my cause to prove that a colony exists or did exist on Mars. I have included this information so that the reader will understand that there is an entire scenario concerning Mars that is separate from my story. Those who are interested can investigate "Mission to Mars" or *Alternative 3* for themselves. It is interesting to note, however, that the documentary entitled "Alternative 3" was shown on a San Francisco TV station sometime around the late '70's. A story has proliferated since that time that the FCC threatened to revoke the station's broadcasting license if it were to be shown again. It wasn't shown again.

The directors of the Montauk Project knew there was a colony on Mars. It is more than likely that they were a part of the conspiracy.

Mars was interesting to the Montauk researchers because they had realized that there was an old technology there. They knew somebody had built the pyramids and face on Mars. These were not natural formations.

According to the information that my associates and I have dug up, the people who were living on the surface of Mars could not get to the underground area beneath the pyramid. The entrances were either sealed over or simply couldn't be found. In fact, it appeared that the big pyramid was sealed better than the pyramid at Giza. Despite all the expensive and fancy technology that was available, the pyramid could not be penetrated.

---

* *Alternative 3* was originally published in the United Kingdom. The first printing in the U.S. was in 1979 by Avon Books, a Division of the Hearst Corporation, 959 Eighth Avenue, New York, New York, 10019.

The scientists at Montauk decided the best approach would be to project right into the center of the Martian underground. The newly discovered Montauk technology gave them the wherewithal to use a spatial warp to get inside. They wanted to get into the underground caverns. These were thought to be set up and administered by a very old civilization.

The time portal took the risk out of the operation as we could look through it. We had a set up with TV monitors so that whatever Duncan would visualize would appear on the monitors. This provided a visual of present time on planet Mars. In order to find the underground area, we kept moving the open end of the vortex until a corridor appeared. At that point, we had Duncan solidify the portal. The away team was then able to walk from Montauk to Mars and be underground.

By this time, Duncan was no longer required to be continuously in the chair. We had learned to have Duncan generate functions with the computer storing them and continually spitting them back. The computer could generally run the transmitter for a short time and had enough memory to modify the flow of time for about four hours. If Duncan would not return after that time, the thought forms being generated would drop out of reality. In such a case, the thought forms would have to be reconstructed from scratch.

The system definitely needed a live being initially. He would create the time portals and hold them open through concentration. Once the opening had been made, we could record on tape what the live person was generating. The tape could then be used on its own to create another opening.

The system was continually refined and improved. If Duncan made a time connection once, it would then be recorded on tape. Because he sometimes had difficulty getting a connection, the tape made it easier and automatic. An entire library was eventually accumulated so that they didn't have to rely on Duncan. It was this development that enabled Duncan to be sent through the vortices himself. This occurred in '82 and '83. He was eventually selected for the team that went to Mars.

Using the time portals, Mars had been scoured for live inhabitants. Researchers had to push back about 125,000 years before they could find any. I don't know what they found out or what they did with the information. Duncan has tried to access this information, but it is deeply buried and difficult to contact.

My personal view is that the pyramid on Mars serves as an antenna. Perhaps there is technology inside of the pyramid. According to Duncan's recollections, he travelled to the inside of the pyramid. He saw technology being operated there and called it "The Solar System Defense". According to his account, the Montauk researchers wanted this shut off. It had to be shut off before anything else could be done. This defense has been shut off retroactive to 1943 which is commonly considered amongst many UFO buffs to be the beginning of the massive UFO phenomena.

There's not much more I can say about Mars at this point except that the movie *Total Recall* is fancifully based upon some of the events that occurred with the Montauk Project. The way they used the chair in that movie is strikingly similar.

Time research continued and countless missions were run until August 12, 1983. This was when the actual lock was made back to 1943 and 1963.

Editor's Note (from Peter Moon)

Ancient historical writings corroborate the idea of an "Elder Race" that was the parent culture of Earth. This subject is addressed in the books "The Black Sun" and "The Montauk Book of the Living" by Peter Moon.

15 ENCOUNTER WITH THE BEAST

On August 5th, 1983, we were given a directive to run the transmitter non-stop – just turn it on and let it go continuously. We followed the orders, but nothing out of the ordinary occurred until August 12th. Then, something very strange happened. All of a sudden, the equipment appeared to drop into synch with something else. We didn't know what function the system was now attuned to, but at that point, the *USS Eldridge* (the ship used for the Philadelphia Experiment) appeared through the portal. We had locked up with the *Eldridge*.

I'm not sure if this was a pure accident, but if the Montauk researchers were trying to hook up with the *Eldridge*, the attempt had to be made on this exact date. This is due to the 20 year biorhythms of the planet Earth (which was a discovery made in the process of these experiments) and the *Eldridge* experiment having occurred on August 12, 1943.

At this point, the Duncan from 1943 appeared and could be seen through the time portal along with his own brother. Both were crew members of the *USS Eldridge*. We kept the Duncan of 1983 from seeing himself so as to avoid a time paradox and resultant negative effects.

The project had now reached apocalyptic proportions. Natural laws were being violated, and it seemed everyone involved felt uncomfortable. Three colleagues and myself had been privately voicing misgivings about the project over a period of months. We had talked about the pitfalls of dealing with time and how this might affect the karma of the planet. We hoped the project would truncate itself.

Consequently, our little cabal created a contingency program that only Duncan could activate. It was designed to crash the entire project.

We finally decided we'd had enough of the whole experiment. The contingency program was activated by someone approaching Duncan while he was in the chair and simply whispering, "The time is now".

At this moment, he let loose a monster from his subconscious, and the transmitter actually portrayed a hairy monster. It was big, hairy, hungry and nasty. But it didn't appear underground in the null point. It showed up somewhere on the base. It would eat anything it could find, and it smashed everything in sight. Several different people saw it, but almost everyone described a different beast. It was either 9 feet tall or 30 feet tall, depending on who saw it. I personally believe it was about 9 or 10 feet in height. Fright does strange things to people, and no one was sure what the exact physical constitution of this monster was. No one was in any frame of mind to calmly and collectively analyze its exact nature.

My supervisor had ordered us to shut off the generators in order to stop whatever type of phenomena was occurring. This didn't work, so it was decided that the thing had to be stopped.

It was decided that the transmitter had to be shut down. There were two efforts made in this direction. One was to send somebody back and turn off the transmitters on the *Eldridge*. They would be smashed if that was what was necessary to shut them down.

The other effort was by myself and the director of the project. We unsuccessfully attempted to shut the transmitter at Montauk. We then went into the power station and disconnected the base from the Long Island Lighting Company. The power kept going and nothing stopped.

We weren't concerned about the lights. We just wanted to stop the transmitter itself. We decided the next best thing to do was to go into the power station and cut the wires leading into the ground from the big transformers. I put an acetylene torch on my back and cut the wires going into the ground. I had to be careful because they were hot. Still, nothing happened. The lights at the base stayed on.

I figured there must have been another power feed somewhere. We went over to the transformer farm next to the transmitter building and cut the wires coming up out of the ground. At that point, the lights at the base went out and the computer stopped. But, the lights in the transmitter building stayed on!

We went into the building and pulled the wires out of the panel that controlled the transmitter; then wires from the transmitter itself. The lights in the building went off, but the transmitter stayed on.

I then went upstairs and cut the actual equipment apart. I cut the conduits. I cut the cabinets. Finally, I cut enough apart that the transmitter just groaned and stopped. All the lights went off. We'd done it. Today, you can still see the torch marks where I'd cut things apart.

It was at this point that the beast stopped moving and faded back into the ether. The portal closed and that was the end of that episode.

After we stopped the transmitter and things settled, we figured out what had happened. When we had first thrown the switches in the power station, none of the lights went off at the base. There was no power coming into the base. When I cut the lines going to the transmitter building, the rest of the base went off, including the computers. The transmitters, however, ran without the computers.

The system had actually gone into a free energy mode. The two systems (i.e. the two generators – one in 1943 aboard the *Eldridge*, one at Montauk in 1983) were locked together. There was a tremendous amount of energy bouncing between the generators. With so much energy between them, all the electrical circuits that were connected remained active. The lights stayed on.

More importantly, the generators established a connection from 1983 to 1943. By bouncing energy between the two time periods, a stable vortex was created. This served as an anchor. Using this vortex, a time tunnel could then be projected to a specific point in time.

For example, if one wanted to go from 1983 to 1993, the '83 to '43 vortex would first have to be functioning to serve as the anchor. The projection to '93 (or whatever other point in time chosen) would come out the '83 end of the vortex.

If one wanted to go to 1923, one would project through the 1943 end of the vortex. Times between 1943 and 1983 could be reached by going through either end of the vortex. Dates after 1963 were accessed through 1983 and dates prior to 1963 via 1943.

This is not to say that all time travel would have to be done in this manner (using the master vortex from 1943 to 1983). During these experiments, no generators were found in either the past or the future that could link up and establish a vortex of this nature. There are, of course, plenty of generators around, but a successful link had to be made. That link required a "witness effect".

"Witness" is an occult term. As a noun, it refers to an object that is connected or related to someone or something. For example, a lock of someone's hair or a picture could serve as a witness. As a verb, "witness" means to use an object to enter a person's consciousness or otherwise have an effect on them.

One example of a "witness effect" would be for someone to take a lock of hair, use it with a love potion, and have the owner of the hair fall in love.

With the Montauk Project, there were three "witness effects". They could be considered as three different levels of witnessing.

The first level consisted of physical people who were actually on the *USS Eldridge*. Any surviving crew members they could find were brought to Montauk for the experiment in 1983. This also included personnel who were considered to be reincarnated since the Philadelphia Experiment. Duncan and Al Bielek were both there and were two of the primary witnesses.

The second level of witnessing concerned technology. The zero time reference generator (referred to previously as the whirligig) used aboard the *Eldridge* was also used at Montauk. When the *Eldridge* was eventually decommissioned in 1946, the whirligig was placed in storage. It was eventually brought out to Montauk and incorporated with the system there. In addition to the whirligig, there were two very strange radio transceivers[*] linking the two projects. They were "cross time" transmitters. They could transmit across time, and they used that to lock up the two projects.

The third level of witnessing was the planetary biorhythm. The term "biorhythm" is esoteric and refers to the higher order channels that regulate life in an

---

[*] I was able to acquire a few of the transceivers that were used in the experiments. Up to this date, I do not fully understand them or their function. It is impossible to get any literature or manuals on the subject. The only possible way to get information on these transceivers is to ask people who have used them. The only answers I've gotten thus far is that they were a highly classified piece of equipment. People I spoke to knew they were for the stealth airplanes, but they didn't know what exactly they were for.

organism. Biorhythms are a result of the resonance upon which nature operates. In humans, the processes of sleeping and eating would involve biorhythms. Of course, there are many subtle ones that could be studied, ad infinitum. When viewing the Earth as an organism, there are also biorhythms. The seasons and the daily spin of the planet would involve biorhythms. The scientists at Montauk exhaustively studied the biorhythms of the Earth and how they related to the entire universe. They discovered that there is a major planetary biorhythm that peaks out every twenty years.

The Philadelphia Experiment occurred in 1943. Although 1983 was forty years later, it was a multiple of twenty and served as a potent witness. It enabled the two projects to link up. I should also mention that it is entirely possible that the link could have been made without the witness effect, however, its application proved very helpful to the project.

The reader should now have some idea of the general theories and applications that were used at Montauk.

After the bizarre occurrences of August 12, 1983, the Montauk base virtually emptied. The power was restored, but lights were left on with everything in disarray. Most of the personnel were eventually rounded up, debriefed and brainwashed accordingly.

DEVASTATED BUILDING

According to legend, this is a building that the beast demolished.
It is to the south of the main base.

THE BEAST

This photo was taken in 1986, well after the
Montauk Project climaxed. It appears to be a giant beast,
however, there was no such beast when the shot was taken.
This would be a phantom phenomenon of some sort, barring a more
natural explanation. The structure is an underground bunker.
It is approximately twelve feet in height.

ENLARGEMENT

An enlargement of the photo on the previous page.
The original print, if looked at with a magnifying glass,
does show what appears to be a snout, eyes and mouth. Unfortunately,
it was a distant shot so the blow up is not of good quality.

16 THE NATURE OF TIME

This book will give rise to many questions, particularly about the nature of time itself. From my experience in talking to groups, I will try to clarify some points that often confuse people.

First of all, the past and the future can be changed.

It will help to consider the idea of a chessboard. In chess, there might be thirty moves in a game. Each one of these moves will create a different layout of the chess board. If one were to "go back in time" and change a move that had been made, it would consequently change all the other board layouts subsequent to that change.

Time could be considered a hypnotic pulse that we all subconsciously agree or submit to. When someone is able to manipulate a change in time, they are also manipulating our subconscious considerations and experiences. Therefore, if time is changed, one wouldn't necessarily realize it.

This scenario implies that we are merely pieces on a grand chess board. To a degree, this is true. For example, retired generals often complain of having been the pawns of international bankers. It is a farfetched comment, but perhaps there would be no war if generals could be truly clued in on the real machinations behind international politics.

There is also the example of Homer's *Iliad* which tells the story of the Trojan War. According to that legend, the gods literally manipulated the characters on Earth like a chess board. The story is ripe with intrigue between mortals and gods. The plots become so intricate and thick that it sometimes seems Homer is trying to provide us with a microcosmic view of the entire universe.

Whatever the case, we are all players in the game that is known as "time". The obvious way to protect one's interest is, of course, to gain knowledge about time itself. Whether one wants to do this by pursuing meditation or astrophysics is an entirely personal matter.

At Montauk, the scientists also viewed the future. The viewers they had gave them the ability to look at multiple futures. Once they chose a particular scenario and activated it by someone or something travelling to it, that future would become fixed. That point would be locked to the time from whence the connection was made. It would create a loop that was fixed.

For example, let's say multiple futures were viewed with different people becoming president. Suppose the future with "Sam Jones" as president was chosen by the researchers, for whatever reason. Linking a person or item from the present would lock in the President Sam Jones scenario no matter what. However, none of this means that a fixed point scenario couldn't be further changed by the scientists doing more manipulations.

At this writing, we are currently in a time loop. This loop extends from where the Montauk researchers penetrated into the past up to where they penetrated into the future. It's fixed and would appear unalterable. However, this does not mean that we are all relegated to being hopeless slaves of time manipulators. The subconscious has its automatic or hypnotic levels, but it also contains the seeds of freedom: dreams. If one can dream something, it can be brought into being.

It is very easy to get philosophical about all of this and get lost in that process. My point with this book is that there has been manipulation of time. This has also exploited individuals and caused untold misery. It could easily be considered to be the work of dark forces.

There is still one major question. Who was really behind the Montauk Project? There are countless intrigues and scenarios one can envision. Religionists can bring in God and the Devil. UFO aficionados can offer a grand scheme of aliens vying for our solar system. Left wingers will offer explanations concerning the CIA and secret government.

I believe that all of the above can shed light on what actually happened at Montauk. It is also my hope that this book will bring more people out of the woodwork. Thus, we can have more answers and less mystery.

PLANETARY CROSS POINT

A traffic circle at the Montauk Air Force Base. To the left is the mess hall and to the right is a dormitory. Within the traffic circle is a planetary grid cross point. Normally, a grid refers to a network of uniformly spaced horizontal and perpendicular lines. In esoteric studies, a grid refers to an intelligent geometric pattern. Theoretically, the Earth and its energies are organized in such a system. If tapped properly, these grids could supply free energy to the world. Dating back to World War I, most military bases have such a cross point, which is usually indicated by a circle around it.

17 THE MONTAUK BASE IS SEALED

After the events of August 12, 1983, the Montauk Air Force Base was abandoned. By the end of that year, there was no knowledge of anyone being on the base.

In May or June of 1984, a crack squad of Black Berets were sent to the base. I believe they were Marines, but I'm not absolutely sure. They were reportedly ordered to shoot anything that moved. Their purpose was to purge anyone who might be on the base.

There was a second team that followed the Black Berets. They removed secret equipment which was considered too sensitive to leave behind.

The next step was to prepare the underground to be sealed. Certain incriminating evidence was removed at this point. I've heard that a room with hundreds of skeletons was cleared out during this evolution.

About six months later, a caravan of cement mixers appeared on the base. Many people saw these trucks. They filled the vast underground areas at Montauk with cement. This included dumping cement down the elevator shafts as well.

The gates were locked up and the base was abandoned for good.

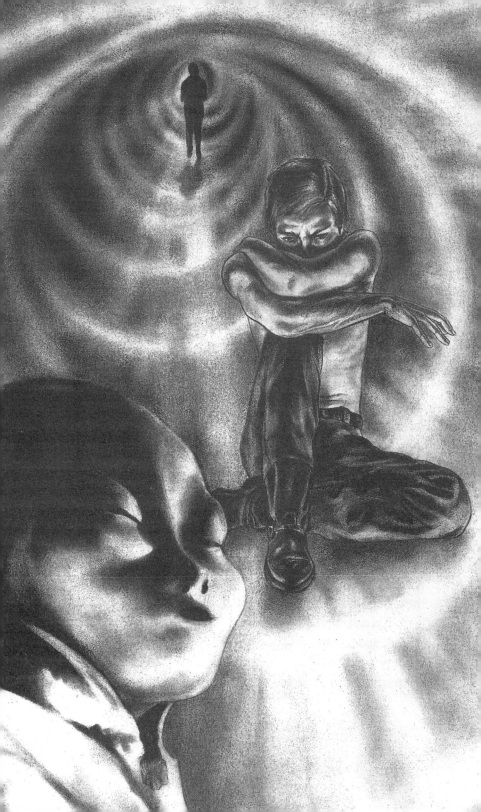

18 MONTAUK TODAY

If one travels to Montauk Point today and parks in the state parking lot near the light house, it is possible to get a good view of the giant radar reflector that sits atop the transmitter building.

For those who are either brave or foolish, one can follow the dirt roads that lead to the base. Most of the entrance gates have been bent or otherwise vandalized so that entrance is easy. This was probably done by local juveniles who sometimes get drunk and have beer parties on the base. However, walking on the base is prohibited by New York State park rangers who periodically patrol the area.* There are also occupied buildings on the main roads to the base.

It should be noted that I am not writing this information to lure people to the base. People are going to be curious after reading this book, and it is my responsibility to warn them. I'm not absolutely sure of the legal technicalities, but walking on the base is probably illegal. One goes at one's own risk.

There are also other dangers to consider.

Two people I know who participated in the Montauk Project visited the area in the late '80s. They claimed they were abducted and do not totally remember what happened to them.

---

* The entirety of Fort Hero, including the inner Montauk Base has since been donated to New York State as a park. While there are peculiar political arrangements concerning the base to this day, the rangers are not out of bounds in keeping people off the grounds. The buildings are in a state of disrepair and are potentially dangerous to those going on a casual walk.

Another person has reported in August of 1991 that video cameras can now be seen from the top of the transmitter building. This is a new development and is rather odd considering it is a vacant and derelict facility. There are also reports that the underground areas of the base are being reopened. This is speculative but should serve to warn anyone making travel plans to Montauk.

EDITOR'S NOTE (from Peter Moon):

Since "The Montauk Project" was published, Camp Hero has drastically changed as a result of this book. You can read about these changes in subsequent publications of Sky Books, particularly "Pyramids of Montauk", as well as the chronological developments as reported in the "Montauk Pulse". Camp Hero is now open to the public.

19 VON NEUMANN ALIVE!

After completing the first draft of this book, a new development took place. It concerns occurrences that started years ago but only came to a resolution recently. This concerns John von Neumann and corroborates the theory that he didn't die in 1958 as is generally believed.

In 1983, I was contacted by a friend of mine in upstate New York who I will refer to as Klark. He knew I was interested in communications equipment and told me about an old time surplus dealer who I will call Dr. Rinehart.* Rinehart was a legend in the local surplus community.

Klark said that the man had a collection of equipment that went back as far as the 1930s and 1940s. A meeting was arranged with Dr. Rinehart on the pretext that I was interested in buying him out. Klark introduced me, and Rinehart showed me his collection on a silver platter. He sincerely wanted to sell out, but I thought his price was too high. A great deal of the equipment was junk and would have cost just as much to cart away.

I considered his prices exorbitant and thought he might be a bit crazy. Apparently, he went even crazier after he met me. Klark visited him again on his own and was met at the door with a shot gun. Rinehart pointed the gun and told Klark that he didn't want to see that bastard Preston on his property. He didn't want to see Klark, Preston or

---

* Dr. Rinehart is a pseudonym used to protect this individual's privacy.

any of their friends on the property. He said he'd shoot them if they showed up.

Klark tried to calm him down and asked what this was all about. He had no idea why the man was so upset. Rinehart said that Preston had come back and robbed him the night he was last there.

As it turned out, somebody had come to the guy's house, strapped him to the chair, ransacked the house and stole money. It certainly wasn't me, and Klark and I were both confused. Years went by, and I had dismissed the puzzling circumstances concerning Dr. Rinehart.

As my memory of the Montauk Project returned, I suddenly recognized Dr. Rinehart. He was actually John Eric von Neumann, the brains behind the Philadelphia Experiment and Montauk Project!

Many years back, perhaps as early as 1958, von Neumann had been assigned to a "witness relocation" program. He was given a new identity as Dr. Rinehart and slipped into a new role as a surplus dealer upstate. He also remained on call to the authorities that ran the Phoenix and Montauk Projects and would work for them whenever he was needed. Sometimes this was for months at a time.

This man not only looked like von Neumann, his doctorates in mathematics and physics were on the wall and came from Germany. Despite this, he claimed never to have left the United States.

It was also apparent that this gentleman's faculties and memory had been tampered with.

I had talked this over with Al Bielek, and we figured that my presence at von Neumann's house was too much for him. He would have remembered me from Montauk and that probably frightened him and caused him to flip out.

All of this is fascinating in its own right, but my main interest was in a strange looking receiver that he had. It is known as an FRR 24 Receiver. I had noticed it on my original visit, and it was still there. I wasn't about to return to his place in view of his threats, but I'd sent people up there and they indicated the receiver was still there.

Al remembered von Neumann as well and wanted to visit him. In fact, von Neumann as Dr. Rinehart had taken a liking to Al. Hoping to get my hands on his receiver, I drove Al upstate to Rinehart's house.

We weren't sure how to approach him on the receiver. We thought about having me wear a disguise but thought it might be easier to have Al buy the receivers on my behalf.

Al got out of the car and greeted him. I remained in the car in hopes that he would ignore me. It started to rain, so Rinehart told Al that they should go to the trailer on the other side of the property. That's where the equipment was. Rinehart walked right by my car and looked me straight in the face. He was friendly and said I should come along, too. Apparently, Rinehart didn't recognize me. I followed them to the trailer as if nothing had ever happened between us.

Al got the guy talking, and I just listened. Von Neumann didn't come through. He was strictly in the identity of "Dr. Rinehart" as he spoke to us.

When he stopped talking, I told Rinehart that I'd heard he had a very large receiver setup where each receiver fits in a rack by itself.

He said "Oh, that thing! I was gonna keep that. But hell, I'm never gonna use it. I can't even move it. I'm gonna keep it, or I'm gonna sell it."

I asked him how much he wanted for it, and he said he would give it to me for a thousand dollars. I told him that Al and I couldn't afford that kind of money, so he suggested a trade.

Al told me to make an offer so I offered $600 for four racks of the receivers. He said that was a little less than he wanted and that he'd have to think it over. We left on good terms and returned home.

Another meeting was arranged some time later. He said that he wanted hi-fi equipment and would be willing to work out a trade. We dug up some hi-fi material and went up again. He looked at it and practically got tears in his eyes. He was excited to see the material and remembered the people who actually designed much of it.

He apologized and said that he really couldn't use any of the material. He wanted cash. If we sold the material, he said we could come back and get the receivers for cash.

We carted everything back to Long Island again. I was frustrated but wasn't about to give up. I called around and found out that I could sell the material. It was worth $750 to other dealers, and I sold it immediately.

I wanted to get his receivers quickly as he was getting known again by national collectors. They would snap up the receivers if I didn't act soon.

I took $800 and went up to see Dr. Rinehart again. I'd taken along some friends to help me move the equipment. Fortunately, it was a clear day, and the weather wasn't going to interfere with our plans.

Dr. Rinehart came out and was in a friendly mood. I showed him $750, but he said he didn't want any money until he was sure I was satisfied with the receivers. He showed us around the place. We went to look at the receivers, and I was surprised. He had four racks of the equipment, and I had remembered only one. He was willing to let them all go for my offer of $750 which was more than fair on his part. I found him to be quite pleasant. In fact, I was a bit puzzled. Initially, he had wanted $1,200 per rack which meant $4,800 for the entire system. Now, it was seven years later, and he was accepting $750. It is my opinion that he wanted me to have the receivers for some reason. I still don't know exactly what it was.

As I poked around and looked at the receivers, my two friends went to the chicken coop as they were interested in some Western Electric equipment that was stored there. Dr. Rinehart was sitting in a chair not far from the receivers. Suddenly, I noticed that he was no longer Rinehart. He was John von Neumann! He remembered his true identity and began to talk.

He definitely remembered me and told me things that were of a sensitive nature which I am obligated not to repeat. He also said that over the years he had seen that millions of dollars were put away in secret Swiss bank accounts. This money was to be used to compensate many of the workers at Montauk who had suffered as a result of the project. Apparently, when I had visited him years back, some sort of signal alerted the secret group that backed the Montauk Project. He was bound up and robbed the next night, and his secret bank books were missing. He now realized that I had not been involved.

I wasn't able to start moving the receivers out until the next day. It was a big job. I took the receivers out of their racks and broke them up so they could be moved safely. Rinehart was there, too, and he started to fade in and out. First, he was Rinehart, then he'd be von Neumann. It was like a yo-yo. Finally, he settled on von Neumann.

As von Neumann, he said that he had obtained these receivers for a very good reason. They were actually capable of tuning in on either of the two projects: Project Rainbow (the Philadelphia Experiment) or the Montauk Project. Further, the receivers were capable of tuning in on the projects from any other space and time in our universe. He also believed this receiver was the main witness from

Montauk to the *USS Eldridge*. He said it could pick up the pattern of the *Eldridge* back in 1943.

It seemed that von Neumann had completed what he had to say. Rinehart returned, and I loaded up the receivers to take back to Long Island.

I wasn't sure how the receivers worked or what they were all about. My first step was to ask Duncan to do a psychic reading. He indicated that the receiver was capable of tuning to any particular point in time by way of zero time. He said that if we could figure out how to tune it, we could tune in to any other point in time.

We realized what von Neumann had already told me: this equipment was a key part of the Montauk time machine. I don't think the particular equipment that I had was on either the *Eldridge* or at Montauk. I think it had been used at the Philadelphia Navy Yard in the 1940s.

I wanted to trace this equipment further to see if it had a logical point of origin. I called the biggest old time radio surplus dealer in the country. They had never heard of the FRR 24 Receiver. I talked to lots of friends in the surplus business and found only one person who had ever seen one or heard of one. This person said that the receiver came out of RCA. He had owned a piece of one of the receivers at one time. It had gone out of his hands when an old man from upstate New York came and paid an exorbitant price for the piece he had of the receiver.

Tracing this back to Dr. Rinehart, he verified that he was the purchaser of the equipment. But, he said that this only accounted for pieces from two of the racks he'd sold to me. There were four in total, and he had to buy the other two racks of receivers from somebody else. I tracked the other person down with Dr. Rinehart's help. This person was a young man who also said the FRR 24 receiver had come from RCA.

I decided to find out how many of these receivers had actually been released. I called up the Surplus Disposal Agency, gave them the number for the receiver, and they did a computer read out. A lady at the agency said that only three FRR 24's had ever been released. All the other systems were either still in use or had been destroyed.

Then, she indicated that, until recently, this receiver had been classified. She said that if any of them had been scrapped, the manuals for them had to be destroyed.

There was also a note indicating that each FRR 24 unit contained seventy-five pounds of silver. The units had reportedly been scrapped and sold to dealers for silver recovery. When scrapped, they are not useful as they would have been put through a crusher.

The report indicated that FRR 24's were only released when the government agreed to sell them to a world communications company. There were three such instances listed. One FRR 24 went to RCA, one went to ITT on the West Coast and another to Vero Beach, Florida.

I tried to trace down people who had actually worked with the FRR 24. Finally, I located a retired gentleman who had worked at RCA Rocky Point (on the eastern end of Long Island). He had worked at the receiver station at Rocky Point.

The gentleman indicated that the FRR 24's had been at the RCA receiving station for years. He raved about the receivers and said they were beautiful and fantastic. When they were turned on, however, he said that a very strange type of interference was picked up all across Long Island Sound. It was a mystery and neither he nor anyone else could figure it out. He also mentioned that the receivers made strange audio noises and that RCA finally decided not to use them.

This was interesting because von Neumann told me that two receiver racks from RCA Rocky Point had been sent back to the 1930s. One ended up in the Philadelphia Naval Yard and was used to track the Rainbow Project in 1943. The other receiver rack ended up at RCA for disassembly and study so that it could be replicated and applied to the technology of the day.

It is interesting to note that in the 1930s, RCA made tremendous strides in radio technology. The years 1933 and 1934 were particularly ripe with new discoveries.

If von Neumann is right, RCA received and analyzed a rack of receivers from the future. It is likely that von Neumann would have sent them back himself.

The receiver rack that ended up at the Philadelphia Naval Yard eventually came into my possession, and I still have it. The disassembled rack was enhanced and improved upon by RCA, and these are what ended up at Rocky Point. This was accomplished through a time loop, thus there are some differences in the RCA receivers (the FRR 24's that I recently got from von Neumann) and the one used during the Philadelphia Experiment. Both receivers, however, have more similarities than differences.

In addition to the above, I also have equipment developed by Tesla that appears to be inspired by the FRR 24's that were sent back from the future by von Neumann.

Whatever the case, radio technology was pushed way ahead in the 1930s. As an engineer and professional radio man, it is my personal conclusion that it couldn't have been done without some major help from somewhere. For example, Nikola Tesla was always upsetting the status quo by saying that he experienced communication with aliens.

There is another major point of interest concerning the FRR 24 receiver. When I purchased them from Rinehart, I noticed that the aluminum housing was corroded

128

on the outside. Aluminum itself does not corrode, but it does when it is mixed with impurities. The aluminum panel in the chassis, however, had no such corrosion. The chassis were therefore made out of a very pure aluminum. Commercial grade aluminum used in radio equipment is usually not that pure.

What does all this tell us?

There had to be a reason that the aluminum was so pure. Recently, it has come to the surface in conventional scientific circles that aluminum can be made into a superconductor. A friend of mine at NASA has told me that mixing mercury with aluminum and alcohol makes micro channels big enough for electrons to channel through the aluminum. This essentially creates a room temperature superconductor.

Rinehart also warned me about the chassis. He said their might be a little bit of mercury contamination on the chassis. Further inspection has revealed that the chassis had some sort of mercury treatment process.

It is currently my belief that the treatment process is related to the silver resonators which are the capacitors and coils. The mercury and aluminum create micro fine channels along the superconducting chassis and the channels become a multi-dimensional resonator.

To conclude, this receiver is actually a multi-dimensional space-time resonator and would be part and parcel of any time machine that was used for the Philadelphia Experiment or at Montauk.

EDITOR'S NOTE: (from Peter Moon)

Per the book "The Prisoner's Dilemma", President Eisenhower announced just before von Neumann died of cancer, that "Johnny will be with us for a long time."

In Memoriam

Preston B. Nichols

———————————

May 24, 1945 — October 5, 2018

In October 2018, as the Silver Anniversary edition of **The Montauk Project** was being prepared for publication, it was learned that Preston B. Nichols passed away at the age of 73. On an extremely hot day in July, Preston had suffered a heart attack and was hospitalized. He was fitted with a pacemaker and was surprised to find out that he had actually designed the electronics for it. Preston then suffered a stroke in September; and while there was some optimism that he might make a full recovery, he took a sudden and unexpected turn for the worse after his room was switched in a rehab center. He could no longer speak and had lost use of his arms and hands. Preston passed away soon thereafter and did not suffer unduly.

On the 27th of October, a memorial service was conducted on the grounds of the Montauk Air Force Station, attended by his friends and admirers. It will be posted on the internet for all to see. Preston was fondly remembered and appreciated for his many contributions.

We miss Preston and thank him for his role in our lives and wish him many blessings on his future journeys.

Blessed Be

A SCIENTIFIC ANALYSIS OF THE RADIOSONDE

(Note: This analysis is not expected to be readily understood by the general lay public. It is included only for those who are technically inclined. It also serves as corroboration of my statement that the government had the means to affect the weather.)

The Radiosonde consists of two variable resistance type sensors. One registers temperature, the other humidity. The temperature sensor is a thermistor where the electrical resistance varies inversely with the temperature. The humidity sensor is an electrolytic resistor where the electrical resistance varies directly with the relative humidity. In most of the Radiosondes, the pressure sensor is of the pressure responsive selector switch type (Baro switch). In essence, the transmitter sees a varying resistance which is selected alternately by the Baro switch or a sequential switcher. A short circuit is selected occasionally which is called the reference mode. This is what the sensors do on the surface. It is also the line the Government releases to the public. Although cursory investigation will show this description to be true, there is also other activity which is secret. The temperature sensor is a carbon bar with precious metals added and acts as an antenna to the DOR function. It also inverts the transform known as DOR energetics. This item is packed in a small vial and has to be installed on fasten stock clips on the arms of the Radiosonde. In order to get a true temperature reading, it is painted white to reflect the sun's radiant heat and sits above the package in the open. This placement is understandable from conventional science, but it cannot be understood from the viewpoint of relativistic sciences

The humidity sensor is an electrolytic resistor. We do not understand its operation because the usual electrolytic resistor varies inversely with the relative humidity. This humidity sensor consists of a grid of conductive lines with an unknown chemical overlay. It acts as an antenna for orgone in-phase. It is also similar to the electrolytic detectors that have been introduced for the detection of esoteric energies. The humidity sensor is also hermetically sealed in a small vial and has to be put in its holder on top of the Radiosonde, thus totally covering and protecting it from direct rain but allowing air to circulate around it. This follows the released line of information.

In later Radiosondes, the Baro switch is replaced with a clock-work driven scanning switch with the addition of a receiver that the Government claims is used as a transponder to track direction and height. This suggests that the Baro switch gives height information which can be read from the pressure but depends on a uniform pressure gradient which our atmosphere does not exhibit. This follows the released information but is grossly inaccurate.

I do not believe that this is the real objective for the Baro switch. In fact, an entirely different purpose is intimated. It appears that the Baro switch is the correlated function which would be necessary to synchronize the DOR busting to the environment of the Earth. It is also apparent that the receiver synchronizes the DOR busting to the environment. At this point, I do not fully understand the sensor scheme.

In addition to the sensors, the other part of the Radiosonde is the transmitter. It is pulse time modulated and the repetition rate of the pulse varies with the resistance presented to the transmitter. There are two types of pulse modulation used. One is where the modulation pulses off the CW (CW = carrier wave) carrier oscillator. The other is where a high voltage pulse supplies B+ (B+ stands for "B batteries" which refer to plate voltage) to the carrier oscillator. There are two frequencies used: 400 MHz and 1680 MHz. The 400 MHz oscillator is comprised of tuned lines with the triode tube in the field of the lines. The 1680 MHz oscillator is of the integral cavity type with the triode tube inside the fields of the cavity.

In the down pulsed CW oscillator transmitter, there are two sections: the modulation oscillator and the carrier oscillator. The modulation oscillator (see page 139) is what generates the pulse which is a triode oscillator with a blocking network in the grid circuit.

The operation of the transmitter is simple. When the oscillator runs, the grid bias builds up across the cap (C) and when the voltage reaches the tube cutoff, the oscillation stops. At this point, C discharges until the tube starts up again. As the grid voltage across C cycles up and down, the oscillator starts and stops; hence modulating the voltage drop across Rp, which is bypassed. The value of C and Rg and Rext and Rref determines the repetition rate of the pulse across Rp. The signal with the pulse and its potentials are capacitively coupled to the carrier oscillator.

This is the usual explanation, but let us consider the relativistic activity. When the tube is cut off, the higher order signal builds up inside the tube like the charge on a capacitor. The longer the tube is cut off, the more relativistic signal builds up in the tube. When the tube turns on and oscillates, it slews between saturation and cutoff, and two things happen. First, the stored relativistic charge is forced out. Second, the slewing of the oscillation between saturation and cutoff has the effect of amplifying the higher order components by "0" point activity. The result is that the signal is amplified and outputs in pulses. From this point, the modulation signal is capacitively coupled to the grid of the carrier oscillator where the pulse stops the oscillation.

When we consider the carrier oscillator (see page 140), the circuit is a standard one. How it was optimized probably has something to do with the placement of the tube in the field of the resonant network and the design of the tube. The higher order operation in the carrier oscillator is similar to the modulation oscillator. When the tube slews from saturation to cutoff, the 0 point of the vacuum is ripped. This results in relativistic gain as well as forcing all signals stored in the tube to the output and the antenna.

The bypassed output of the modulation oscillator, which is pulsed potential (scalar) at approximately 7 MHz, is coupled to the grid of the carrier oscillator and slews the Q point* from saturation and cutoff. The "0" point activity sends out bursts of relativistic signal which replicates closely the input signal from the sensors.

The pulse modulator transmitter uses a delay line pulse modulator with a thyratron, charging reactor, blocking diode, pulse forming network, and a pulse transformer that generates 1400 V pulses which drives the carrier oscillator. The thyratron is fired off by the output of the same modulation oscillator as in all of the rest. The modulation oscillator loads the relativistic signal into the delay line through the thyratron which is cut off but still has "0" point gain. When the thyratron fires, everything in the pulse forming network is loaded into the carrier oscillator tube as a 1400 V pulse which rips apart the vacuum and results in a high "relativistic gain" through the usual "0" point activity.

The whole package is operated by a battery pack which lasts approximately three hours.

* "Q point" means quiescent point. This refers to the point where the tube rests.

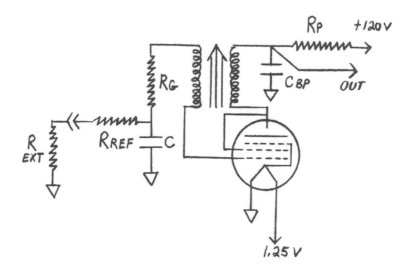

MODULATION OSCILLATOR

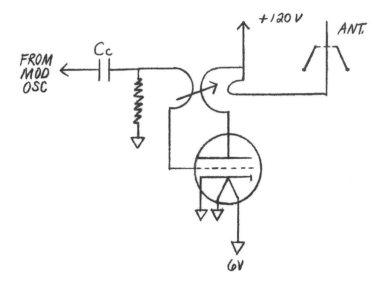

CARRIER OCSILLATOR

B WILHELM REICH

Although the Government had great appreciation for the discoveries and developments of Wilhelm Reich, it appears they had very little use for him personally. He was under pressure from the AMA and FDA for years on charges of quackery. Eventually, he was given a very stiff prison sentence for contempt of court when he refused to appear. The subsequent burning of his books and destruction of his equipment may be unparalleled in modern times for its outrageousness.

His claims about jamming the drives of UFOs didn't win him too many friends either. He concluded that the drives of UFOs ran on cosmic or orgone energy. He developed a "space gun" on the basis of orgone theory and was able to make UFOs fade away with regularity according to eyewitness accounts.

After he was in prison, authorities reportedly gave him express permission to work on anti-gravity equations. This is odd to say the least; especially if they believed he was a quack.

Whatever the exact facts and details of Wilhelm Reich are, it appears that he was used for his inventive genius and then put away so he couldn't disseminate his discoveries elsewhere. The systematic condemnation of his work only backs this up.

(See Editor's Note from Peter Moon on next page)

Another more secretive aspect to Reich's research that was used at Montauk had to do with hidden or repressed sexuality in humans. All of this equated to the idea that repressed life force causes an individual to "misfire" and shows up in their "body armor" or physical structure. Reich's techniques were designed to release such repressions by physically addressing the body armor along with whatever verbal therapy was deemed appropriate. It does not require a stretch of the imagination to realize that his techniques and philosophy might create a lot of outrage in certain quarters of society. The most outrage, however, came from the Government itself.

Long before Reich was arrested and put in prison, his unorthodox techniques were being studied by secret factions within the Government, and they were compared to and blended with the most sophisticated clinical mind control techniques of Josef Mengele, known as the Angel of Death in the Nazi death camps. These techniques, reminiscent of the mind control techniques of Hassan Sabbah's Assassin cult of the Middle Ages, equate to a deep "sexual hypnosis" induced in the consciousness of the subject. The long and short of this is that these techniques were used to program people to carry out whatever orders were given.

These secret or lost techniques of Reich were discovered in an old barn in Rangely, Maine when Preston was searching for and buying old radio equipment, a hobby for which he was well known for. These lost techniques are not a part of what typical Reich enthusiasts would know about.

Reich was eventually killed in prison, and the easiest reason to believe why is that his knowledge would have either prevented or combatted the ill intentions that were being used against the people.

C
MIND CONTROL AND THE PERSIAN GULF WAR

I was still laid off when the Persian Gulf War erupted and had the opportunity to watch the live briefings. Live briefings are interesting to me because information sometimes comes through that would otherwise be edited out.

In one report, a CNN reporter said that he had just returned from Kuwait where he had travelled with an American patrol. They had noticed a patrol of about thirty Iraqis on the next sand dune. While the Americans were wondering how they would get the Iraqis to surrender, a U.S. helicopter suddenly appeared and flew over the Iraqis. By the time the helicopter had reached the next sand dune, the Iraqis had their hands up and were surrendering.

This is all very suspicious in itself. These were the same Iraqis who fought a Holy War against Iran for eight years.

The next news of interest that I noted came towards the end of the conflict when Brigadier General Neil was questioned by a British reporter from the BBC. The reporter asked the General about his plans to get the Iraqi soldiers out of the deep bunkers that the Germans built for the Iraqis. These bunkers were known to be extremely well fortified, and it was a good question.

General Neil said, "We bring in the psychological..."*

He then broke his sentence with coughing. It didn't sound like a real cough but that he had caught himself

---
* The quotations by General Neil are not exact quotations but are paraphrases based upon my memory of the actual events.

saying something he shouldn't reveal. When he stopped coughing, he continued speaking.

"I'm sorry, we bring in the helicopters with PA (public address) systems and we talk 'em out."

To me, his statements were very significant. It is apparent to me that the General had made a mistake and had to continue his sentence in the same vein. In my opinion, he was going to say something like "psychological broadcasting helicopters". He had helicopters on his mind and in order to make it less obvious, I believe he changed what he was going to say and referred to "PA system helicopters".

I did some research into the Iraqi bunkers and discovered that the Americans had been trying to get the plans for their construction from the East Germans. The Americans wanted to know how to penetrate the bunkers. They did retrieve the plans and found out the walls were very thick. Even after the air blitz, the Iraqis were still deep in the bunkers. They had electricity, entertainment and enough food and water to last at least six months. The bunkers were three feet thick and probably could have withstood a nuclear blast. The Iraqis also had the equipment to tunnel out if necessary.

The British reporter knew it wouldn't be easy to get the Iraqis out of the bunkers. That's why he asked the question. I believe it is absurd to suggest that these fanatic soldiers would have surrendered under the mere threat of PA system helicopters.

D NIKOLA TESLA

Nikola Tesla was born in 1856 in what is today known as Yugoslavia. Known as the "Father of Radio", he was clairvoyant and had different paranormal abilities. Most notable was his vision as a youngster that he would build an alternating current generator that would revolutionize the way that Mankind harnessed electricity.

Tesla received a renaissance education and learned to speak several languages. He worked his way across Europe as an inventor and electronics engineer. In Paris, his genius came to the attention of one of Thomas Edison's associates and Tesla was invited to meet the famous inventor. Although Edison hired him, the two never got along well.

Edison's utilities used direct current which required a power plant every few miles or so. Tesla tried to convince him that alternating current was more effective and less costly to operate. Edison was stubborn and Tesla's brilliance must have made him feel insecure. Here was a man whose genius was far superior to that of Edison's!

Edison would never back up Tesla's plans to revolutionize the world with alternating current. The two finally had a complete falling out when Tesla advised Edison that he could upgrade his entire facility by building new machines and replacing the old ones. Edison offered him $50,000 to complete the task. Tesla designed twenty-four types of machines and effectively enhanced the factory. Edison was very impressed but wouldn't pay the money. He claimed it was just his "American sense of humor."

George Westinghouse was an inventor himself and recognized the genius in Tesla. He backed Tesla's plan to harness alternating current from Niagara Falls and the world has never been the same since. Meanwhile, Edison tried to prove that alternating current could be deadly to humans and went to the extent of electrocuting a dog in public (with alternating current) to prove his point. Edison ended up embarrassed and humiliated.

Tesla's career went on the rise and his experiments were of vast renown. He demonstrated remote control with small boats in Madison Square Garden, but many people dismissed it as witchcraft.

He even generated lighting between the Earth and sky at Colorado Springs. This experiment was particularly remarkable because he put light bulbs to the raw ground and they lit up. This demonstrated that the Earth's surface was a conductor of electricity. This proved that if the proper vehicles were used that the entire population of Earth could enjoy free energy.

Tesla created a huge tower on Long Island and sought to build a system that would provide free energy. While in progress, the financier J.P. Morgan pulled the rug out from under Tesla. He didn't want free energy.

Tesla's career went into a decline and his reputation was hurt. Part of this was caused by his periodic admissions that he received communication from aliens. His receivers supposedly picked up transmissions from Mars.

No one ever denied he was an electronics genius, but because he understood supernatural phenomena, he was held in suspicion. Today, many of my engineering colleagues consider him a "nut" who just happened to be brilliant in electronics. This is a very convenient explanation.

It is my opinion that he was incredibly ahead of his time.

E

THE PHILADELPHIA EXPERIMENT AND ITS RECONCILIATION WITH THE MONTAUK PROJECT

In 1912, a mathematician named David Hilbert developed several different methods of new math. One of these was known as "Hilbert Space". With this he developed equations for multiple realities and multiple spaces. He met Dr. John von Neumann in 1926 and shared his information. Von Neumann took a lot of the systems he learned from Hilbert and ran with it. According to Einstein, von Neumann was the most brilliant of mathematicians. He had an uncanny ability to take abstract theoretical concepts in math and apply them to physical situations. Von Neumann developed all kinds of new systems and math.

A Dr. Levinson had come along and developed the "Levinson Time Equations". He published three books which are now very obscure and almost impossible to find. An associate of mine did dig up two of them at Princeton's Institute for Advanced Study. All of this work was to serve as a background for the invisibility project which would apply the theoretical principles to a large hard object.

Serious research into the subject of invisibility began in earnest in the early 1930s at the University of Chicago. Dr. John Hutchinson Sr. served as Dean at this particular time and was privy to the work of Dr. Kurtenhauer, an Austrian physicist then at the University. They were later joined by Nikola Tesla. Together, they studied the nature of relativity and invisibility.

In 1933, the Institute for Advanced Study was formed at Princeton University. This included Albert Einstein and John von Neumann, a brilliant mathematician and scientist. The invisibility project was transferred to Princeton shortly thereafter.

In 1936, the project was expanded and Tesla was made the director of the group. With Tesla on board, partial invisibility was achieved before the end of the year. Research went on to 1940 when a full test was done in the Brooklyn Naval Yard. It was a small test with no one on board the vehicle. The ship used was powered by generators from other ships, connected by cables.

Another scientist, T. Townsend Brown, became involved at this point. He was known for his practical ability to apply theoretical physics. Brown had a background in gravity and magnetic mines. He had developed counter measures to the mines with a technique known as degaussing. This would trip the mines at a safe distance.

There was a big brain drain on Europe in the 1930s. Many Jewish and Nazi scientists were smuggled into the country. Much of this influx has been attributed to A. Duncan Cameron Sr. Although we know he had extensive connections, his exact relationship to intelligence circles is still a mystery.

By 1941, Tesla had full confidence of the powers that be (FDR). A ship was procured on his behalf, and he had coils wrapped around the entire ship. His famous Tesla coils were also employed on the ship. However, he grew wary because, as the project developed, he knew there would be problems with personnel. Perhaps he knew this due to his ability to fully visualize his inventions in his mind. In any case, Tesla knew that the mental state and bodies of the crew would be affected severely. He wanted more time to perfect the experiment.

Von Neumann disagreed with this vehemently at the time and the two never got along. Von Neumann was a brilliant scientist but did not embrace metaphysics for its own sake. Metaphysics was old hat to Tesla, and he had built a successful legacy of inventions based upon his unique prescience.

Part of what made his views so controversial was that, during his experiments in Colorado Springs, circa 1900, he said that off planet intelligence had contacted him via consistent signal messages when Mars approached. This also occurred in 1926 when he had radio towers erected in the Waldorf Astoria and at his New York City lab. He claimed to receive information that he'd lose people if things were not changed. He needed time to design new equipment.

Tesla's requests for more time were not heeded. The Government had a war to win and additional time was not granted. Tesla went through the motions but secretly sabotaged the operation in March 1942. He was either fired or quit. He is supposed to have died in 1943, but there is arguable evidence to suggest he was whisked off to England. A look-alike derelict is supposed to have been put in his place for the funeral. He was cremated the day after his body was found which was not in keeping with the tradition of his family's Orthodox faith. Whether or not he died is controversial. That secret papers were removed from his safe has never been in question.

Von Neumann was named director of the project. He did a study and determined that two huge generators would be required for the experiment. The keel for the *USS Eldridge* was laid in July of 1942. Tests were done at dry dock. Then, in late '42, von Neumann decided that the experiment could be fatal to people, just as Tesla had suggested. Ironically, he still got upset at the mention of Tesla's name. He decided a third generator would do the trick. He had time to build one but never got the third one to synchronize with the other two. It never worked because the gear box was incompatible. The experiment went out of control and a Navy technician was zapped, went comatose for four months and left the project. They pulled out the third generator. Von Neumann wasn't satisfied, but his superiors weren't going to wait any longer.

On July 20, 1943 they decided it was ready and made tests. Duncan Cameron Jr. and his brother, Edward, were in the control room to operate it. The ship was no longer at anchor and orders came by radio to turn it on. Fifteen minutes of invisibility ensued. There were immediate problems with the people. They got sick, some experiencing nausea. There were also mental illnesses and psychological disorientation. They needed more time, but a final deadline was given for August 12th, 1943. The orders came from the Chief of Naval Operations, and he said he was only concerned with the war.

Trying to avoid damage to individuals involved, von Neumann tried to modify the equipment so that only radar invisibility would be achieved, not literal sight invisibility.

Six days before the final test on the *Eldridge*, three UFOs appeared over the ship.

The switch was thrown for the final test on August 12th, 1943. Two of the UFOs left the area. One was sucked up into hyperspace and ended up in the underground facility at Montauk.

Reports from Duncan indicated that he and his brother knew things were going to go wrong with the August 12th experiment. However, for three to six minutes, things looked good. It appeared it might work without any devastating effects. They could see the outline of the ship — it

146

hadn't disappeared. Suddenly, there was a blue flash and everything was gone. There were problems. The principal radio mast and the transmitter were broken. People were jammed in the bulkheads. Others were walking around in an insane state.

Duncan and Edward Cameron did not suffer the same trauma as their shipmates. They had been shielded in the generator room which was surrounded by steel bulkheads. The steel acted as a shield to the RF energy. As they witnessed things falling apart, they tried to shut off the generator and transceivers but were unsuccessful.

At the same "time", another experiment was going on forty years later at Montauk. Research had revealed that the Earth, like humans, has a biorhythm. These biorhythms peak out every twenty years on August 12th. This coincided with 1983 and provided an additional function for the connecting links through the Earth's field for the *Eldridge* to be pulled into hyperspace.

The Cameron brothers could not turn off the equipment on the *Eldridge* because it was all linked through time to the generator at Montauk. They figured it wasn't safe to remain on the ship and decided the best alternative would be to jump overboard in hopes of escaping the electromagnetic field of the ship.

They jumped and found themselves pulled through a time tunnel and onto dry ground at Montauk on August 12th, '83 at night. They were found quickly and taken downstairs.

Von Neumann met Duncan and Edward and indicated he knew they were coming. He was now an old man. He said that there had been a lock up in hyperspace and that he'd been waiting since 1943 for this date. He told the time travellers that the technicians at Montauk were unable to shut things down. Duncan and Edward were required to go back to 1943 and shut the generator off. von Neumann even told them that the historical records showed that they had turned it off. But they hadn't done it yet! He told them to destroy any equipment if that's what it took.

Before returning to 1943 for good, Duncan and Edward did some missions for the Montauk group. They made a number of trips back to 1943. On one of these trips, Duncan passed through the time portal and entered the time tunnel. Duncan somehow entered a side tunnel and got caught there. Side tunnels were a mystery and remain so. Even though the Montauk scientists theoretically considered side tunnels non-existent, Duncan was warned not to enter them if they should appear. Edward soon ended up in the same tunnel with Duncan.

A group of aliens revealed themselves. Apparently, the side tunnel was an artificial reality created by the aliens. They wanted a piece of equipment before they would let their captives go. This equipment was a very sensitive instrument that charged the crystal drive to the UFO that was underground at Montauk. The aliens didn't seem to mind leaving a ship, but they were very intent on keeping the drive source a mystery to humans.

Duncan and Edward returned to Montauk and retrieved the drive for the aliens. Eventually, they were able to return to the *Eldridge* and carry out von Neumann's orders. They smashed the generators, transmitters and cut every cable they could find. The ship finally returned to its original point at the Philadelphia Naval Yard.

Before the portal closed, Duncan returned to Montauk in 1983. His brother, Edward, remained in 1943. Duncan is not sure why he returned. It has been suggested that he may have been under orders or programmed to do so.

This adventure turned out to be a disaster for Duncan. His time references totally dissolved, and he lost his link to the time line. When time references are lost, one of three things happens: aging slows down, remains the same or speeds up. In this case, it speeded up. Duncan began to age rapidly. After a short amount of time, he began to die of extreme old age.

We're not sure how this happened, but we believe von Neumann transferred him to another time. Scientists were enlisted to help him. They couldn't let the Duncan from 1943 die. He was not only invaluable to the project, he was elaborately involved with the entire scope of time. His death could have created bizarre paradoxes and had to be avoided.

Unfortunately, Duncan's body was dying and there was nothing that could be done to alter the rapid aging. But, there was another alternative. Research had already demonstrated that each human being has their own unique electromagnetic identity. This was commonly referred to as one's "electromagnetic signature" or just "signature". If this "signature" could be preserved when Duncan's body ceased to function, it could theoretically be transferred to a new body.

The Montauk scientists were already intensely familiar with all of Duncan's electromagnetic manifestations from the exhaustive research that had been done. By some means, I'm not sure how, his "soul" or "signature" was transferred to a new body.

They sought help from one of their most loyal and effective agents: A. Duncan Cameron Sr., who happened to be the father of Duncan and Edward Cameron.

Duncan Sr. was a mysterious character. He was married five times over the course of his life. He had numerous connections and didn't seem to work. He spent his time building sailboats and travelling to Europe. Some have alleged that he smuggled Nazi and/or German scientists into the U.S. via his boating activities.

There is practically only one tangible piece of evidence that connects him to intelligence circles. He appeared in a photo of a special graduation for intelligence personnel at the Coast Guard Academy. He was not officially affiliated with the Coast Guard in any way.

Through the use of the Montauk time techniques, the Montauk group contacted Duncan Sr. in 1947. They informed him of the situation and told him to get busy and have another son. He now had a different wife than Duncan Jr.'s original mother. Duncan Sr. cooperated and a child was born, but it was a girl. His directions were to produce a son. Finally, a boy was born in 1951. "Duncan" was chosen as the name for this child, and this is the same Duncan I know today.

The Montauk techniques are obviously remarkable, but they were not sophisticated enough to move Duncan from 1983 straight back to 1951. There could have been other factors involved, but it appears the scientists had to rely on and use the twenty year biorhythms of the Earth. As Duncan's original body was dying, he was transferred to 1963 and "installed" into the new body provided by Duncan Sr. and his wife.

Duncan Jr. has no memories prior to 1963. It is also obvious that whoever occupied his body between 1951 and 1963 was forced out.

I have often heard accounts of a secret project that was run by ITT at Brentwood, Long Island in 1963. It is entirely possible that transferring Duncan to a new body was the focal point or a very important part of this project. Whatever the circumstances, this project would certainly have been trying to somehow utilize the Earth's biorhythm that occurs every twenty years.

Edward Cameron had returned to 1943. Duncan was in 1963.

After the August 1943 experiment, the Navy brass didn't know what to do. Four days worth of meetings ensued with no conclusions. They decided to do one more test.

In late October of 1943, the *Eldridge* disembarked for the final experiment. No personnel were to remain on board. The crew boarded another vessel and controlled the equipment on the *Eldridge* remotely. The ship became invisible for about fifteen or twenty minutes. When they boarded it, some of the equipment was missing. Two transmitters and a generator were gone. The control room was a burnt shambles, but the zero time reference generator was left intact. It was put into secret storage.

The Navy washed their hands of the entire operation and officially launched the *USS Eldridge* with its official record. The ship was eventually sold to the Greek navy who later uncovered the log books and found that everything before January of 1944 had been omitted from the records.

According to Al Bielek's account, Edward Cameron continued his career in the Navy. He had top level security clearance and probed into many sensitive areas such as "free energy" vehicles and devices. He was outspoken and complained about improper procedures. For whatever reason, he was brainwashed to forget the Philadelphia Experiment and anything else to do with secret technology.

Al has stated that age regression techniques were used to put Edward Cameron into a new body in the Bielek family. The Bielek family was chosen as there was only one child in the family and that baby had died by the time of his first birthday. Edward was substituted and the parents were brainwashed accordingly. Edward has since been known as "Al Bielek".

Age regression techniques have been traced back to Tesla. When he was working on the original Philadelphia Experiment, he developed a device to help sailors in the event that they lost their time locks. The purpose of this device was to reestablish an individual with his normal time locks in the event he had been disoriented from time travel. The Government or someone allegedly used this Tesla device and developed it for physical time regression.

Tesla said that if the time locks of an individual are moved ahead in time, one could actually remove age. If one's time locks were pushed twenty years younger, the body would be referenced to those locks.

Edward Cameron now became Al Bielek. Al grew up with his own identity and education and became an engineer. Eventually, he ended up working at Montauk. It was not until the mid 1980s that Al began to get memories from his earlier identity. To this day, he continues to doggedly research the Philadelphia Experiment and is planning to write another book. He intends to prove, even to the most skeptical, that the Philadelphia Experiment did indeed happen.

F QUANTUM LEVELS OF EXISTENCE
(According to Preston Nichols)

When I am referring to quantum levels of existence, "quantum" refers to the different or many possible levels. "Quantum" comes from the Latin root "quantis" which means quantity.

Understanding multiple realities is key to understanding time. Conventional physics does not deny the possibility of parallel existences, but it is mostly concerned with theories about matter and anti-matter. As there is much evidence to prompt scientific investigation into the area, there are currently about sixty theories around the world and ten in the U.S. that deal with quantum levels of reality.

I am offering my own theory based upon my experiences, some of which are covered in this book. Like any proper scientific theory, this is being offered because it has proven workable to me in the laboratory. It will also help the reader get a better grasp of how time functions.

What exactly is a parallel reality?

It would be a world or universe that has almost everything we have here. If we switched into it, we'd see another body that would represent us in the other existence. The parallel universe would not necessarily behave exactly like the one we're familiar to. It would have unique properties unto itself.

It is my understanding that we exist in a number of parallel realities. We are primarily conscious of "our reality" because we are focused or referenced to it. The parallel universes might reach our consciousness through dreams, ESP, meditation or artificially induced mental states.

It is now important to consider the overall view and what these different realities might look like in schematic form. Einstein theorized that if one travels in a straight line from any particular point in space that one would eventually end up in exactly the same place that they started at. This could be considered a full loop. We are not going to go into the equations of that, but the general reader can grasp this by understanding what Einstein called a time toroid. For this purpose, a toroid can be likened to a two dimensional doughnut. Einstein likened the entire universe to a time toroid. He theorized that if one started in a straight line from any given point on the outside of the doughnut, that one would end up at the exact opposite end of the doughnut. Both of these points would be essentially the same, except that one could be considered "positive"

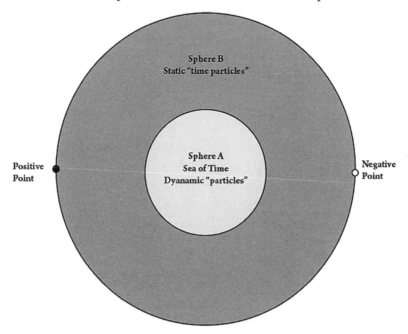

and the other "negative". As they are both points in the infinite stream of time, one point could be called "positive infinity" and the other "negative infinity".

For purposes of explaining my theory, I have extrapolated upon Einstein's idea and have converted his time toroid into a three dimensional sphere. The reader can easily picture a smaller sphere inside of a bigger sphere. For clarification, we will call the inside sphere, Sphere A, which can be likened to a softball. The bigger sphere will be called Sphere B and can be likened to a basketball (which is hollow).

Our experiences in different realities all occur within the realm of Sphere A. If you start from a point on Sphere A and travel in a straight line, you will eventually come back to where you started from.

Within Sphere A, everything is dynamic and moving. It is time as we conceive it.

The area between Sphere A and Sphere B is not dynamic at all. In fact, it is considered to be at rest. We can postulate this area to be a sea of time particles. These are not particles in any ordinary sense. In fact, trying to describe them in this three dimensional example is awkward. We are simply assuming these non-moving particles of time exist because we can sense them (if only in idea form). These particles between Sphere A and Sphere B would be the same as the particles in dynamic time (within Sphere A) except that they are static, i.e. at rest.

Incidentally, we are not conscious of static time because our "normal" reality is built upon dynamic functions or dynamic time.

A reality in time is created when God or someone creates a stress on the wall of Sphere A. This stress will cause the dynamic particles inside of Sphere A to move and travel through the sphere until there is a loop, thus completing the Alpha and Omega (beginning and end).

Our reality can be considered one gigantic loop. It could have started with the big bang or the beginning of the universe and would eventually end there, but it would in fact continue ad infinitum.

When someone or something then takes that loop that we are in and creates a new time stress and changes reality, a new loop is created that is, in fact, an alternate reality. The original loop cannot be obliterated or denied. It will still be there. The new loop could be modified in any way the modifier chooses. It could be an opening in 1963 that goes to 1983. Everything in between those times would be an alternate loop. It would not be an entire loop of its own, but would be added to the original loop of our normal reality. In this way, partial loops would be added on to our original time line, and we can call this conglomeration of loops a manifold. Each loop could also be called a manifold (a manifold generally refers to something that has many parts).

As different alternate realities are created off of the original loop, additional manifolds are added to the sphere and make it swell. In addition to the alternate realities created by changing the reality of a particular time line, there could be parallel realities created in the beginning of time that also have their own "original" loops as well. There is an infinite number of loops and manifolds possible.

Some people may wonder about Sphere B in the above example. It is basically there to make the theory fit. At this point, I can add no further significance to Sphere B except that it serves as a wall containing time stress particles. It could possibly be part of a bigger scheme of metaphysics.

Now that you have a grasp of how these time loops and manifolds fit into an overall picture of the universe, there is another key question that must be asked. Is it possible to gain consciousness of the other loops or manifolds?

Yes, it is. This is what happened on my roof when I was putting up my Delta T antenna (as discussed in Chapter Six). That antenna has a subtle interdimensional effect on the nature of time itself. It enabled me to regain consciousness of an alternate time line that I had been put on against my innate will.

It is therefore possible for others to travel from one time loop to another. In fact, it appears that this is the entire reason for the Philadelphia Experiment and Montauk Project in the first place. This theory indicates that not only was an alternate time loop created but that this loop enabled a vast influx of alien UFOs to come to this planet. UFOs have always been around, but there is no denying the sudden frequency of reports in the 1940s.

Even if you do not accept any of this as the truth, it is quite obvious that this is the type of advantage an alien race might have over us.

The next point I want to address is that parallel realities are based upon principles common to electromagnetics. For example, it is common knowledge that alternating current is created by an alternating difference in potentials. This is best demonstrated in a coil, where the current and potential are shown in the following diagram.

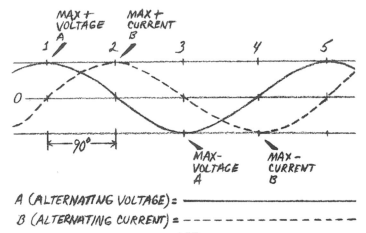

NOTE ON PRECEDING DIAGRAM: (A and B are said to be ninety degrees out of phase. A "cycle" of current/ voltage from peak to peak is 360 degrees. 90 degrees out of phase means that when alternating voltage A is at its peak strength, alternating current B is at zero.)

It is the relationship between the voltage and the current that makes alternating current function. Further, the relationship between the current and voltage is inextricable.

Likewise, to draw a metaphysical analogy, our reality is represented by the "A" wave in the previous diagram while "B" would be a parallel reality. Just as there is an interplay between the voltage and current, there is one between two different realities.

Extrapolating from these principles, it can be understood that parallel realities are ninety degrees out of phase with our "normal reality". In other words, if there is a parallel reality, one has to consider that it is has potential energy. It is not activated of itself. It would also be ninety degrees out of kilter from our normal viewpoint. The fact that it is potential energy means that it has the capability to flow to our reality and vice versa.

This explains that there is not only a relationship between electromagnetic principles and other universes but suggests that, by utilizing electromagnetic principles, one can theoretically enter the realm of other realities. These would include the alternate time loops of which I have already spoken.

It is hoped that the above will give the reader a general understanding of how electromagnetic principles were used to manipulate time at Montauk.

GLOSSARY

amplitron – A high powered UHF amplifier. At Montauk, this served as the final amplifier of the transmitter before a function was radiated out the antenna. A large tube, it weighed 300 pounds and measured 35 inches in its largest dimension.

biorhythm – This is an esoteric term and refers to any regularly repeating life function in an organism. A biorhythm is perhaps best understood in terms of Oriental "Ki" or "Chi" energy which is the life force that regulates the entire body. Acupuncture addresses biorhythms in order to affect a cure. When the planet is considered as an organism, biorhythms would include all the subtle functions that make life possible and regulate it. The seasons, rotation of the Earth and spinning of the galaxy would all be taken into account. Legendary places such as Stonehenge are considered to be constructed in harmony with the biorhythms of the planet.

cathode – In a vacuum tube, the electron emitting material is called a cathode. In an electrolytic cell, it is the negative electrode from which current flows. In essence, it is a source of flow.

cycle – A unit of activity within a wave that continually repeats itself. A cycle will go up and down before it repeats itself. If you visualize ocean waves that are all uniform, the series of waves would be called the "wave". The one ocean wave that a surfer might ride would be a "cycle".

Delta T – Short for "Delta Time". Delta is used in science to indicate change, thus "Delta T" would indicate a change in time

Delta T antenna – An octahedronal antenna structure that is designed to bend time. Visually, it looks like two pyramids sharing the same base. By definition, it can actually facilitate shifting time zones. Two coils are placed vertically around the edges of the pyramid structure at 90° angles to one another. A third coil surrounds the base. Shifting time zones was accomplished by pulsing and powering the Delta T antenna, as is discussed in Chapter 12. Even when the antenna is not powered, it has a subtle interdimensional effect on the nature of time itself.

DOR — Stands for "Dead ORgone" (see definition of "orgone"). This refers to life energy that has become stagnant or negative. DOR could be considered the antithesis of life energy.

"electromagnetic bottle" – This refers to a "bottle effect" that is created when a specific space is surrounded by an electromagnetic field. The specific space itself is the inside of the "bottle". The walls would be the electromagnetic field. When people or objects are within the specific space, they would be within an "electromagnetic bottle".

electromagnetic wave — When an electric charge occurs that oscillates (swings back and forth), a field around the charge is generated. This field is both electric and magnetic in nature. This field also oscillates which in turn propagates a wave through space. This wave is called an electromagnetic wave.

frequency – the number of waves or cycles per second.

Helmholtz coils – Commonly, Helmholtz coils refer to two identical coils that are separated by a distance of one radius of the coils themselves. (You can visualize this if you think of two hula hoops parallel to each other.) When the coils are electrified, they produce a homogeneous magnetic field over a larger volume of space than does a single coil.

Hertz – (abbr Hz) This is simply one cycle of a wave. A wave consists of numerous cycles that are repetitions of one cycle. To be a bit more technical, hertz is the complete fluctuation of a wave from plus (the highest point) to minus (the lowest point). Five hertz would be five cycles per second.

MHz – Megahertz, which are equivalent to 1,000,000 hertz.

non-hertzian component – This term does not exist in conventional science. It refers to the etheric component of electromagnetic waves. Theoretically, the non—hertzian component is a wave function. Instead of oscillating transversely, it oscillates with the direction of propagation which is known as longitudinal (i.e. sound waves). It could be looked at as an "acoustical" electromagnetic wave.

orgone – This refers to life energy or sexual energy as observed by Dr. Wilhelm Reich. It is the positive energy that "makes us tick".

oscillator – A device that establishes and maintains oscillations. To oscillate means to swing back and forth. In electronics, an oscillation refers to a regular variation between maximum and minimum values, such as current or voltage.

phase – The time interval between when one thing occurs and the instant a second related thing takes place.

phase conjugation – This is the process whereby a wave comes back from a received source that is an imaginary image of a transmitted wave. In other words, when a radio wave is transmitted, an image goes back to the transmitter by the process of conjugation. (For more information, one can read up on modern electro-optics theory.)

Phoenix Project – A secret project that commenced in the late 1940s. It researched the use of orgone energy, particularly in regards to weather control. It eventually inherited the Rainbow Project and included the Montauk Project itself. "Phoenix" was an official code name.

psycho-active – This pertains to any activity or function that has an affect on the mind or psyche. In this book, psycho-active refers primarily to any electromagnetic function or electronic equipment that influences human thinking and behavior.

Psychotronics – The science and discipline of how life functions. It includes the study of how technology interacts with the human mind, spirit and body. Science, mathematics, philosophy, metaphysics and esoteric studies are united through the study of psychotronics. It would also include other realities and how we interface with other dimensions of existence.

pulse modulations – These are sent as a series of short pulses which are separated by relatively long stretches of time with no signal being transmitted.

RF – Radio Frequency. Frequencies above 20,000 hertz are called radio frequencies because they are useful in radio transmissions.

radio wave – An electromagnetic wave that carries intelligent information (pictures, sound, etc).

relativistic – Relativistic functions refer to activities that are out of our normal reference frame. It also concerns how activities in other reference frames relate to ours. Relativity embraces the concept of everything without any limitations, including other dimensions and the entire universe(s).

sideband – This is the component of radio waves that actually carries the intelligent information.

space-time – When you study higher level physics, it becomes apparent that space and time are inextricably related to each other. It is considered less accurate to refer to just space or time by itself (because they don't exist by themselves). That would be like saying your mouth ate the dinner.

time reference(s) — This refers to the electromagnetic factors by which we are connected to the physical universe and the stream of time. Consciousness of time can be likened to a deep hypnosis which causes one to be in sympathy with the various frequencies and pulses of the physical universe.

transceiver — An instrument that serves as both a receiver and a transmitter.

transmitter — A device or unit that sends a signal or message.

wave – A state of motion that rises and falls periodically is called a wave. It can be transmitted from one particular area to another with no actual transport of matter taking place. A wave consists of many cycles and can carry signals, pictures or sounds.

.

Author's Note

Since the original publication of *The Montauk Project: Experiments in Time* in 1992, I have encountered and learned far more than I bargained for than when I originally engaged Preston Nichols in order to write the original book. The most significant and spectacular of these encounters occurred on August 11, 1999 when Dr. David Anderson, a time control scientist, walked into my life and began to attend our monthly "Montauk Night" meetings we once held on Long Island. His company, the Time Travel Research Center, had achieved the ability to actually slow time down or speed it up within the boundary of a small self-contained field. This was originally funded by the medical field who recognized the value of such capability in order to preserve organs for transplants. Since that time, Dr. Anderson's technical capabilities have expanded to an extraordinary degree, and this includes the capacity to safely send and return individuals, objects or technical devices into the past or future.

While the amazing developments of this technology have spanned over two decades, Dr. Anderson has remained a very elusive and mysterious character, but he has nevertheless maintained a consistent yet intermittent presence in my life. My adventures with him are chronicled in my quarterly newsletter, the *Montauk Pulse*, and this includes him sponsoring me to fly to Romania in 2008 where I have developed a relationship with "Department Zero", the most secretive and powerful faction of that country's secret service. Their activities include the discovery of the most astounding find in the history of archeology: an ancient chamber beneath the Romanian Sphinx in the Bucegi Mountains, the focal point of which includes a data bank that is rendered in holographic visualizations. Not only is there a Projection Hall where one can holograhically review world history, there are three tunnels which lead deep into the Inner Earth that connect to a network of similar installations across the globe. These adventures are chronicled in the Transylvania Series, the first one of which is *Transylvania Sunrise*. Written by Romanian author, Radu Cinamar, I have personally edited the English language version of these books, all of which are published by Sky Books. Available in e-book format on Kindle and iTunes, you can find the hard copy editions at our website which is *www.skybooksusa.com*.

I also encourage you to visit the Time Travel Education Center at *www.timetraveleducationcenter.com* where you can sign up and watch seven free videos that clearly demonstrate that time travel does not violate the laws of mathematics and physics. Although rendered at a level that an eighth grader can understand, it might require a few times through the material as you might have to rethink the way you look at the world.

You can also access the original patent application for Dr. David Anderson's Time Reactor™, a device which not only manipulates time but provides vast amounts of inexpensive energy. While such discoveries have been shrouded in mystery and have been met with an incredible resistance, Sky Books stands at the forefront of the unveiling of these mysteries. Thank you for your interest and support.

There is still so much more to explore.

THE ASTONISHING SEQUEL

MONTAUK REVISITED: ADVENTURES IN SYNCHRONICITY pursues the mysteries of time brought to light in The Montauk Project and unmasks the occult forces behind the science and technology used in the Montauk Project. An ornate tapestry is revealed which interweaves the mysterious associations of the Cameron clan with the genesis of American rocketry and the magick of Aleister Crowley and Jack Parsons. Montauk Revisited carries forward with the Montauk investigation and unleashes a host of incredible characters and new information.

249 PAGES, ILLUSTRATIONS, PHOTOS AND DIAGRAMS........................$19.95

THE ULTIMATE PROOF

PYRAMIDS OF MONTAUK: EXPLORATIONS IN CONSCIOUSNESS awakens the consciousness of humanity to its ancient history and origins through the discovery of pyramids at Montauk and their placement on sacred Native American ground leading to an unprecedented investigation of the mystery schools of Earth and their connection to Egypt, Atlantis, Mars and the star Sirius. An astonishing sequel to the Montauk Project and Montauk Revisited, this chapter of the legend propels us far beyond the adventures of the first two books and stirs the quest for future reality and the end of time as we know it.

256 PAGES, ILLUSTRATIONS, PHOTOS AND DIAGRAMS........................$19.95

THE BLACK SUN

THE BLACK SUN: MONTAUK'S NAZI-TIBETAN CONNECTION explores the intriguing connection between the Montauk Project and the Nazi-Tibetan alliance. This includes the connection to advanced technology at Brookhaven Labs at Yaphank which also boasted the largest contingent of Nazis outside of Germany. Photos are included of the mysterious Vril flying craft build before and during World War II. All of this leads to the Third Reich's quest for holy relics and a penetrating look in the the secret meaning behind the Egyptian and Tibetan "Books of the Dead."

256 pages, ILLUSTRATIONS, PHOTOS AND DIAGRAMS........................$24.95

TRANSYLVANIAN SUNRISE

Transylvanian Sunrise is the story of a remarkable andf unprecedented archeological discovery made in 2003 beneath the Romanian Sphinx in the Bucegi Mountains. Radu Cinamar had the opportunity to visit this secret site where he witnessed a holographic Hall of Records left by an advanced civilization and also three mysterious tunnels leading deep into the bowels of the Inner Earth. *Transylvanian Sunrise* chronicles the political intrigue surrounding the discovery of these modern day artifacts and gives a concise and coherent description of them, the prospect of which represents the dawn of a new era for Mankind. 288 pages, ISBN 978-0-9678162-5-8...$29.95

TRANSYLVANIAN MOONRISE

Transylvanian Moonrise corroborates Radu's story with newspaper articles as he is sought out by a mysterious alchemist who introduces him to a Tibetan Lama. These two take Radu on a mystical journey from Transylvania to Tibet where he receives a secret initiation and a sacred manuscript from the blue goddess Machandi. This is not only a remarkable story, but it is an initiation of the highest order that will take you far beyond your ordinary imagination in order to describe events that have molded the past and will influence the future in the decades ahead. 288 pages, ISBN 978-0-9678162-8-9...$29.95

MYSTERY OF EGYPT

In **MYSTERY OF EGYPT**, Radu is part of an expedition to explore on of the mysterious tunnels in the holographic chamber: the one to Egypt. In this journey, they encounter ancient artifacts that look more futuristic than they do ancient. It is a detailed account of a remarkable adventure that includes further interactions with Cezar Brad, the head of Romanian Intelligence's Department Zero, and Elinor, the enigmatic alchemist. While these claims are more than controversial, Cezar ignites further controversy when he shares some of his experiences since their last meeting. An amazing follow-up to Radu's second book, *Transylvanian Moonrise*, *Mystery of Egypt* also includes explorations in time to the First Century A.D. 240 pages, ISBN 978-1-937859-08-4...$29.95

THE SECRET PARCHMENT

THE SECRET PARCHMENT — FIVE TIBETAN INITIATION TECHNIQUES presents give invaluable techniques for spiritual advancement that came to Radu Cinamar in the form of an ancient manuscript whose presence in the world ignited a series of quantum events, extending from Jupiter's moon Europa and reaching all the way to Antarctica, Mount McKinley and Transylvania. An ancient Romanian legend comes alive as a passage way of solid gold tunnels, extending miles in the Transylvanian underground is revealed to facilitates super-consciousness as well as lead to the nexus of Inner Earth where "All the Worlds Unite."
288 pages, ISBN 978-0-9678162-5-8..$29.95

THE WHITE BAT

THE WHITE BAT — THE ALCHEMY OF WRITING is the story of one of Peter Moon's most mysterious adventures in synchronicity that is centered around his dream of a white bat. Told in a personal narrative, this book synthesizes the dream process with the creative process and teaches you to do the same as it integrates the emergence of ancient Tibetan texts with the remarkable discoveries of a holographic chamber beneath the Romanian Sphinx. The materialization of the white bat heralds the reawakening of a primordial culture the ancients called Hyperborea.
288 pages, ISBN 978-1-937859-15-2..$22.00

INSIDE THE EARTH

INSIDE THE EARTH — THE SECOND TUNNEL continues the incredible adventures of Radu Cinamar as he actually visits different civilizations within the Inner Earth and offers plausible scientific data to backup his stories and various claims. Radu also explains why the Inner Earth has remained so elusive previously and shares a a unique way to actually penetrate the Inner Earth through the process of feeling and the effects that will develop from such an experience. Multiple illustrations are included revealing the geography of Inner Earth. We also meet Radu's old friends Cezar and Dr. Xien who share in the process of educating and initiating humanity with regard to its hidden history.
240 pages, ISBN 978-1-937859-08-4..$29.95

SkyBooks ORDER FORM

We wait for ALL checks to clear before shipping. This includes Priority Mail orders. If you want to speed delivery time, please send a U.S. Money Order or use MasterCard or Visa. Those orders will be shipped right away. Complete this order form and send with payment or credit card information to:
Sky Books, Box 769, Westbury, New York 11590-0104

Name
Address
City
State / Country   *Zip*
Daytime Phone (In case we have a question)   (   )

☐ *This is my first order* ☐ *I have ordered before* ☐ *This is a new address*

Method of Payment: ☐ *Visa* ☐ *MasterCard* ☐ *Money Order* ☐ *Check*

—   —   —

Expiration Date   *Signature*

TITLE	QTY	PRICE
The Montauk Pulse (1 year - no shipping US orders)...$20.00		
Montauk Project PEARL ANNIVERSARY EDITION...$22.00		
Note: There is no additonal shipping for the Montauk Pulse if you are in the United States. *Subtotal*		
For delivery in NY add 8.625% tax		
U.S. Shipping: $5.00 for 1st book plus $1.50for 2nd, etc.		
Foreign shipping: e-mail us at skybooks@yahoo.com		
Total		

Thank you for your order. We appreciate your business.